Environmental Sustainability in Building Design and Construction

Xiaoming Wang • Sayanthan Ramakrishnan

Environmental Sustainability in Building Design and Construction

Xiaoming Wang
Swinburne University of Technology
Melbourne, VIC, Australia

Northwest Institue of Eco-environment
and Resources
Chinese Academy of Science
Lanzhou, Gansu, China

Sayanthan Ramakrishnan
Swinburne University of Technology
Melbourne, VIC, Australia

ISBN 978-3-030-76233-9   ISBN 978-3-030-76231-5   (eBook)
https://doi.org/10.1007/978-3-030-76231-5

This Springer imprint is published by the registered company Springer Nature Switzerland AG
The registered company address is: Gewerbestrasse 11, 6330 Cham, Switzerland

Preface

This book is aimed to serve university students and academics who wish to understand sustainability aspects in building construction for the purpose of learning and practice. The information is presented on an orderly manner on why sustainability in construction is required, how they are being implemented and potential benefits of sustainable construction practices over traditional construction methods. The insights into the negative impacts of current construction methods on resource efficiency, energy and emission as well as climate change are discussed.

Sustainability – as a broader perspective – has received significant attention among researchers, engineers, academics and policymakers in the last four decades. This is mainly due to the alarming rates of resource consumption and the fact that the Earth's resources are limited. The consumption of resources at a higher rate than replenishing can lead to the unavailability of those resources for future generations. In particular, the construction sector is a major contributor to resource consumption due to the increasing demand in construction of buildings and infrastructure. The increasing population growth, high living standards and people spending more time indoors are the primary reasons for increasing demand of buildings.

Apart from the resource consumption rates, the impacts on the environment due to the currently increasing industrial activities are also becoming a major issue. Compared to the CO_2 concentration in pre-industrial age (220 ppm), the current CO_2 concentration is at a higher level (more than 400 ppm) and it continuously increasing. Other changes in natural environment due to anthropogenic effects (i.e. human caused pollution in the environment) are also becoming inevitable. Furthermore, the changes in CO_2 have caused many subsequent consequences such as global temperature increase, sea level rise, and more frequent and high-intensity extreme events. Without reducing the anthropogenic effects, it would be very hard for our society to sustain a liveable environment in the future. Once again, construction sector is one of the major contributors for the adverse impacts on our environment.

The remedy for these concerns lies in acquiring appropriate knowledge on sustainable practices simultaneously while learning building construction methods. This is because the understanding of the negative impacts of current construction practices will support to improve sustainable building design and construction, so that its objectives are fully achieved by following new and efficient approches in construction. Indeed, there are many instances during building construction and operation when we cannot avoid the negative impacts to the environment, but such impacts can be negated or offset by following sustainability practices in the other stages. This book sets out to discuss those aspects.

The first part of the book (Chaps. 1 and 2) provides fundamental information on sustainability, sustainable development and sustainable construction, which helps students to become familiar with new terms in sustainability and also assists in understanding why sustainability remains as a core topic in this sector. The sustainability policies developed internationally and nationally in Australia are also discussed.

The middle part of the book (Chaps. 3, 4, 5, and 6) covers the issues in current construction practices of buildings, occupants and the environment. The quantitative assessment of the concerns in terms of energy and carbon involved in construction is discussed with the aim of understanding life-cycle energy or carbon accounting of buildings. The resource efficiency in building construction, including energy, material and waste, is discussed along with how current construction practices utilise these resources and the possible approaches to follow sustainable practices. The construction sector has been identified as one of the major contributors to carbon emission that leads to climate change, and the changing climate also remains a major threat to the safety and efficiency of buildings. Therefore, the definitions, measurement and projection of climate change aspects are also discussed in this part of the book.

The last part of the book (Chaps. 7 and 8) discusses the implementation methods to achieve sustainable built environment for present and future climate conditions with consideration on the resilience of buildings and infrastructure in changing climate.

The underlying theme of this book is to enhance the understanding of sustainability in construction by students and academics and to incorporate the knowledge in engineering design and construction practices. It must be stressed that the sustainability aspects should be integrated in all stages of construction including, design, construction, operation and disposal of the structures.

Melbourne, VIC, Australia   Xiaoming Wang
                                                                       Sayanthan Ramakrishnan

Contents

Chapter 1
Introduction

1.1 Overview

As indicated by the Fifth Assessment Report of the United Nations Intergovernmental Panel on Climate Change (IPCC AR5), the built environment accounts for 30% of global energy consumption and 19% of energy-related greenhouse gas (GHG) emissions, and it has been projected that this energy use and related emissions may double or potentially even triple by mid-century due to several key trends (Pachauri et al., 2014). The resultant worsening effects of the increased energy use and emission are broadly identified as global warming and related anthropogenic climate change effects causing extreme weather and climate events, including an increase in warm temperature extremes, extreme high sea levels, and the number of heavy precipitation events in a number of regions. In contrast to the reported energy and emission trends, the projected building energy use may stay constant or even decline by mid-century, compared with today's levels, if today's technology, cost-effectiveness, and sustainability are substantially improved and implemented. Thus, built environment serves a global opportunity for the integration of sustainable development with the goal of achieving a low-carbon future.

This book discusses the subject of sustainable develovpment applied to the building construction and built environment area. The issue is complex as it involves resource efficiency, environmental impacts (including climate change), human health, building economics, and social development. Most of these issues are interrelated and must be considered together to obtain an overall understanding. In the past, development decisions and processes were largely made separately in economic, environmental, and social domains. Later on, they are driven either by ecological or by economic aspect with consideration of other requirements, also known as one-pillar models. Since the publication of Brundtland Report in 1987, the notion of sustainable development had been defined with three-pillar models, where development decisions are taken by systematically and equally considering all economic, social, and environmental requirements.

© Springer Nature Switzerland AG 2021
X. Wang, S. Ramakrishnan, *Environmental Sustainability in Building Design and Construction*, https://doi.org/10.1007/978-3-030-76231-5_1

The aim of this book is to encourage the readers to think about these issues independently, logically, and objectively. The book provides more in-depth discussion on the major issues related to built environment such as energy, materials, water, waste, climate change, etc. and outlines various strategies and options for sustainability implementation in these aspects. The book will outline:

- The review of basic concepts of sustainability, sustainable development, and sustainable construction as well as international and local regulation developments in the area of sustainability;
- The understanding of the concept and science of climate change, climate risks, mitigation and adaptation concepts as well as its impacts on buildings;
- The discussion of the attributes of sustainable indicators (i.e., energy, land, water, materials, and waste) in building construction;
- The resilience and adaption of buildings to changing climate; and
- The practical measures that can be implemented in sustainable building design and construction.

1.2 Organization of This Book

This book is organized into eight major chapters followed by the summary and conclusion. The first part of the book, Chaps. 1 and 2, provides students and academics with information that helps to understand basic concepts of sustainability; various definitions and terms of sustainability in building construction are discussed in the first part of the book since the students find it difficult to understand these new terms when they are used in the later part of the book. It also addresses the various sustainability policy developments implemented internationally and nationally in Australia.

The second part of the book, Chaps. 3 to 6, discusses the issues in current construction practices and how these issues can be qualitatively and quantitatively assessed to compare with the issue resolution measures developed as part of the sustainable development measures. The fundamental resources of the built environment including energy, materials, water, and waste are discussed in detail, and the effects on these resources as a result on unsustainable practices are also reported. The identification of the issues in sustainable construction practice is crucial for the students and teachers. This part also summarizes the effect of climate change on the built environment and the methods to project future climate condition using different scenario approaches.

The last part of the book, Chaps. 7 to 8, discusses the implementation methods to achieve sustainable built environment for present and future climate conditions with the consideration of the resilience of buildings and infrastructure as a result of changing climate.

The following section explains the synopsis of each chapter with the main topics discussed in the chapter.

1.2.1 Chapter 1: Introduction

The introductory chapter reviews the background of sustainability and sustainable development with the key concepts of sustainable development in the construction sector. Meanwhile, it also discusses the frameworks used to assess the sustainable development in the construction sector, such as using TBL and life cycle approaches.

1.2.2 Chapter 2: National and International Developments

Sustainability and sustainable development have emerged as an increasingly influential driver of change in public policy, community attitudes, and corporate governance. Around the world, all sectors of society have been developing and interpreting the principles of sustainability according to their specific context (Ashe et al., 2003). This chapter discusses some of the key international developments by organizations such as the United Nations, the Organisation for Economic Co-operation and Development, and the International Organization for Standardization (ISO) as well as national developments by federal, state and territory, and local governments.

1.2.3 Chapter 3: Climate Change and Built Environment

Climate change has now become a major field of research and is considered by many government and institutional bodies as one of the major driving forces behind the sustainable development agenda. This chapter gives fundamental concepts of climate change, climate science, and the causes and impacts of climate change, especially on built environment. This chapter also describes current activities to cope with the changing climate such as climate adaptation, mitigation, and the integration of mitigation and adaptation.

1.2.4 Chapter 4: Energy and Carbon Emission

This chapter focuses on examining the current trends of energy use and its impacts on the environment from a global perspective and in Australian context. It first gives an overview of different energy sources, production, and consumption and then discusses the relationship between energy per capita and human development as a mean of assessing the sustainable development. More importantly, the energy and carbon dioxide (CO_2) emission of building sector during the life cycle of buildings will be demonstrated in detail with the aid of carbon-accounting frameworks applied in construction. Various energy principles and technologies to improve building energy efficiency will also be proposed and discussed.

1.2.5   Chapter 5: Materials and Water

This chapter discusses the use of resources such as materials and water in the construction of buildings as well as its implications to the environment. All products used in building construction require using and transforming raw materials extracted from the environment. Various informed strategies for using material inputs efficiently, such as using recycled and reclaiming of wastes as inputs, will be discussed. Lack of water is another major constraint to industrial and economic growth. The ways to improve water efficiency in buildings will also be discussed.

1.2.6   Chapter 6: Sustainable Waste Management

Traditional approach to waste management relies on the natural environment to absorb and assimilate unwanted by-products. Environmental impacts associated with waste disposal include land contamination, methane emission, odor, toxicity, and consumption of land resources. Different types of waste, disposal methods, associated environmental impacts, and waste management aspects are discussed in this chapter. Furthermore, with the concern of construction and demolition, waste remains as a major contributor to solid waste generation; this chapter will discuss various effective waste minimization management strategies in the construction sector.

1.2.7   Chapter 7: Sustainable Building Design

The purpose of building construction is to improve the social well-being of occupants while having a minimal impact on the environment. This chapter discusses various implementation methods to achieve a sustainable building construction through efficient design, construction, operation, and disposal methods. Three sustainable building concepts of green building design, low-energy design, and zero-energy design are discussed along with the methods and approaches to achieve these concepts. The key strategies discussed in this topic can be very helpful for engineers to implement in their new designs and refurbishment of existing buildings, while it provides prospective thinking to academics and students for developing new technologies to cater these strategies.

1.2.8   Chapter 8: Resilience and Adaptation in Buildings

Despite the fact that energy efficiency in buildings is an important feature of sustainable development, its operative performance to the changing climate should also be ensured. Buildings are designed to operate for longer periods, preferably exceeding 40–50 years, and the initial design of buildings should enable the proactive adaptation to a greater extent. This chapter discusses the threats of changing climate on the serviceability of buildings in terms of occupant thermal comfort, resilience to heat waves, and the durability of building materials. The strategies of resilient design for these events will also be discussed.

1.3   Sustainability and Sustainable Development

The discussion of environmental issues on the built environment is steadily increasing both nationally and internationally. The terms "sustainability," "sustainable development," etc. are now being used widely under many different contexts. The concept of sustainability, however, remains elusive despite the large amount of writing devoted to the subject. "What is sustainability in the context of building?" is the issue to be discussed in this topic.

1.3.1   Sustainability

This term is (Waas et al., 2014):

- An attribute or characteristics of a process or a state of a system;
- A potential capacity of a system to maintain its process and state at a certain level indefinitely; and
- A balance between demand and supply, consistently maintained amid systems that are interconnected and interacted as well.

Figure 1.1 illustrates the five key elements of sustainability including the system or systems to be assessed, its or their states of interest related to sustainability, interactions or processes between systems, time or duration of the system to be maintained, and overall objectives to develop sustainability. Considering earth as our system, our demand on nature perceived as through its current state or through many processes should not exceed the availability of the nature in present. It should also ensure that the resource availability does not become insufficient in the future. If we explain in simple terms, in a sustainable world, society's demand on nature is in balance with the nature's capacity to meet the demand. Meanwhile, the definition of sustainability varies large depending on the consideration in system, process, states, time, and targets. In this context, we provide the definition of sustainability in three different terms of social, economic, and ecological (environmental) aspects.

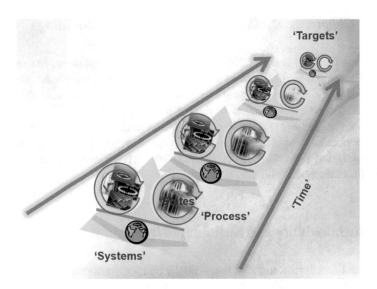

Fig. 1.1 Key elements of sustainability

In the **social term**, sustainability is defined as the way that may improve the social capital through the encouragement of social justices, education, equality, and participative democracy (Parkin & Sharma, 1999). Institutional aspects of social development may also be considered, which is deemed to play a considerable role in implementing strategies. Littig and Griessler (2005) defined social sustainability as a quality of societies, signifying the nature–society relationships, meditated by works, as well as relationships within the society.

In the **economic term**, sustainability can be defined as the way that human manages their economy to preserve its productiveness and prosperity through efficiency, investment, diversification, and balance of internal demand and external supply.

On the **ecological (environmental) term**, sustainability can be defined as the way that human interacts with the biosphere to maintain its life support function on the aspect of biological diversity, ecosystem conservation, and regional interconnectedness. In general, it is often related to the effort to minimize environmental impacts, such as pollutions and the consumption of natural resources. It is often measured by ecological footprint, in terms of consumption of natural resource and generation of waste, describing how much biologically productive land and water an individual or a process requires to produce all the resources it consumes and to absorb the waste it generates. Ecological sustainability leads human settlement and economic development toward a more environmental-friendly way.

It can be found that these definitions of sustainability include some or all of the key elements described above (i.e., system, state, process, time, and target). We request the readers to understand these definitions and identify the key elements of sustainability in these definitions as well as if there are any key elements missing in these definitions (the answers are provided at the end of the chapter). In fact, there are many other aspects of sustainability. It is therefore perceived that sustainability should equally achieve social, economic, and ecological goals, which is generally known as the *TBLs*.

The selection of key indicators in sustainability assessment is important as they should characterize the system, states, process, time, and target of the boundary. Some of the selection criteria of key indicators mentioned by (Fricker, 1998) include:

- Multidimensional, linking two or more in TBLs;
- Forward-looking, capable of application in prediction;
- An emphasis on local wealth, resources, and needs;
- An emphasis on appropriate levels and types of consumption; and
- Easy-to-understand and link to available data.

Meanwhile, the process to select sustainability indicator could be described by Fig. 1.2. In general, the system to be assessed has to be clearly defined before the objectives contributed by sustainability are defined. More subobjectives could be further defined until measurable indicators could be identified and selected.

As sustainability covers broad ranges, the indicators of sustainability are related to multiple disciplines. For example, Fricker (2008) reports the following indicators to measure the urban sustainability:

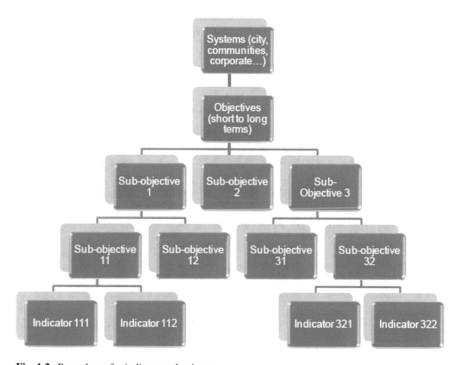

Fig. 1.2 Procedures for indicator selections

- Income per capita;
- Solid waste, water/energy consumption per capita;
- Proportion of workforce in the employ of the top 10 employers;
- The number of good air quality days/years;
- Diversity and population of specified urban fauna;
- Distance travelled on public transport relative to private transport per capita;
- Residential densities relative to public space in inner cities;
- Relative hospital admission rate for selected childhood diseases; and
- Proportion of low birth weights among infants by income groupings.

1.3.2   Sustainable Development

In the past, development decisions and processes were largely made separately in economic, environmental, and social domains, as shown in Fig. 1.3. Later on, they are driven either by ecological or by economic requirements with consideration of other requirements. In particular, the development that gives priority to the ecological requirements is known as one-pillar models (Littig & Griessler, 2005).

Sustainable development is to ensure humanity living within the means of what our planet may provide (Estes, 2010) and avoiding ecological overshooting that leads to resource depleted and eventually used up. Therefore, it has to systematically and equally take into account all economic, social, and environmental requirements, known as three-pillar models (Littig & Griessler, 2005). It needs to consider

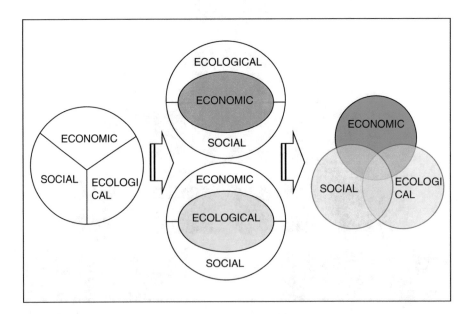

Fig. 1.3 Drivers in the process from a single requirement to sustainable development (Ashe et al., 2003)

a balance amid economic, environmental, and social aspects. Satisfaction of economic and environmental requirements maintains a process or a state of a developed system viable, but not necessarily socially equitable, as shown in Fig. 1.3. There are also multiple pillar models proposed, which can arguably be associated with the three-pillar models.

There are many proposed definitions for sustainable development.

- The generally accepted definition of sustainable development came from the Brundtland Report (WCED, 1987): "development that meets the needs of the present without compromising the ability of future generations to meet their own needs."
- The Australian Commonwealth (Lumley & Armstrong, 2004) have defined the sustainable development as "development that improves the total quality of life, both now and in the future, in a way that maintains the ecological processes on which life depends."
- The Australian Building Codes Board (Board, 2009) has proposed the following definition for sustainable development: "development that effectively balances long and short term economic, environmental and social considerations to meet the needs of the present without compromising the ability of future generations to meet their own needs."

1.3.3   Key Themes in Sustainable Development

Despite the many definitions, there are a number of key themes. The reasons for the many definitions are that no single definition can manage to embrace all the themes and that different emphasis is put on the themes depending on the preferences (or preoccupation) of the groups involved. The key themes involved in the definitions are:

- Concerns that the earth resources are finite and that there is a need for better resource efficiency;
- Concerns that developments may reduce the biodiversity and upset the ecology on which all life depends;
- Concerns for future generation;
- The need to improve quality of life for all;
- The need for equity between different groups of people on earth;
- The need to balance between competing goals (economic, environmental, and social); and
- The realization of the interdependency within and between all communities on earth.

1.3.4   Key Principles of Sustainable Development

Despite the pervasive uncertainty surrounding most issues, a number of key principles have been evolved during the discussion on sustainability. The most fundamental principle of sustainability development is intergenerational equity on the aspect of natural capital that has to be shared equally between now and the future.

Other principle that underlines the sustainability includes (Dresner, 2008; Fricker, 1998):

- The humility principle: It recognizes the limitation of human knowledge;
- The precautionary principle: It advocates caution when in doubt; and
- The reversibility principle: It requires not to make any irreversible change.

In Australia, the principles of sustainability development are described in various intergovernmental agreements, including the following:

- Inter- and intragenerational equity: "the present generation should ensure that the health, diversity and productivity of the environment is maintained or enhanced for the benefit of future generation" (COAG, 1997);
- Precautionary principle: "where there are threats of serious or irreversible environmental damage, lack of full scientific certainty should not be used as a reason for postponing measures to prevent environmental degradation" (ibid., p. 13);
- The TBL approach: "decision making processes to effectively integrate long term and short term economic, environmental and social considerations"; and
- Conservation of biodiversity: "the variety of all life forms—the different plants, animals and micro-organisms, the genes they contain and the eco systems of which they form a part."

While there is a general agreement on these principles, how they are to be implemented is still subject to considerable amount of controversy and debate.

1.3.5   Key Questions in Sustainable Development

The discussion on sustainability usually revolves around the following key questions:

- Can the current world development continue? (Do we have enough resources, food, energy, forests, water, etc. to go on for the foreseeable future? What are the effects of the reduction in forest areas?)
- What will the pollution do to us? (air pollution, acid rain, indoor air pollution, water pollution, and waste)
- What problems are we creating for future generations? (chemical problems, biodiversity, and global warming)

There are no clear-cut answers to the above questions because of the following reasons:

- Complexities of the environmental systems;
- Inadequate science;
- Unreliability of prediction—because of (a) and (b)—not only past predictions have been proven to be wrong, but the future developments of technologies might alleviate or remove the problems altogether;
- Philosophical problem of putting human development outside the "natural eco system," whereas humans are very much part of the system; and
- Nature is always changing, thus, restoration back to which stage remains unclear

1.4   Sustainable Construction

1.4.1   Overview

The construction industry is one of the most dynamic industrial sector in global economy as well as the third largest contributor to gross domestic product (GDP) in the Australian economy. In chain volume terms, the construction industry accounted for 9.0% of GDP in 2019–2020, compared with 6.8% in 2008–2009 (Industry & Committee, 2020). The construction industry operates in both the private and public sectors, engaging in three broad areas of activity, including residential building, nonresidential building, and engineering construction.

However, as indicated by ABS Survey of Energy, Water and Environment management among Australian business in 2008–2009, there is a considerable gap between environmental significance and implementation to address environmental issues in the construction industry, as shown in Table 1.1. The concept of sustainable development in construction should be introduced more in the sector, which is in fact beyond the environmental issues as discussed previously.

Table 1.1 Adoption rate of environmental management by Australian business in 2008–2009

Environment management	Adoption rate (%)
Businesses with some form of environmental plan/policy or system	5.2
Businesses that undertook environmental management activities	27.3
Businesses that undertook water management practices	21.4
Businesses that have conducted energy usage audits	1.9
Businesses that have energy performance targets or indicators	3.6
Businesses that undertook energy efficiency or energy reduction measures	51.1
Businesses that undertook measures to prepare for the proposed Carbon Pollution Reduction Scheme	6.7
Businesses that purchased GreenPower	1.2

Source: ABS

There are just as many definitions for "sustainable construction." The term can be taken as is an abbreviation for sustainable development as applied to construction area. Many attentions have practically been paid to sustainable building construction although they were referred to sustainable construction (Plank, 2008). Some commonly quoted definitions of sustainable construction are:

- "The creation and responsible management of a healthy built environment based on resource efficient and ecological principles" (Kibert, 1994);
- "A way of building which aims at reducing negative health and environmental impacts caused by the construction process or by buildings or by the built-up environment" (Dabirian et al., 2017);
- "In its own processes and products during their service life, aims at minimizing the use of energy and emissions that are harmful for environment and health, and produces relevant information to customers for their decision making" (Team & BRE, 1999);
- "Sustainable construction means that the principles of sustainable development are applied to the comprehensive construction cycle from the extraction and beneficiation of raw materials, through the planning, design and construction of buildings and infrastructure, until their final deconstruction and management of the resultant waste. It is a holistic process aiming to restore and maintain harmony between the natural and built environments, while creating settlements that affirm human dignity and encourage economic equity" (CIB & UNEP-TETC, 2002); and
- "A holistic process aiming to restore and maintain harmony between the natural and built environments, and create settlements that affirm human dignity and encourage economic equity (Du Plessis, 2002)."

A range of stakeholders can be involved in sustainable construction, as shown in Fig. 1.4, including regulators, standard organizations, designers, contractors, product manufacturers, construction industry, maintenance institutes, owners, developers, users, and community. Therefore, participatory development involving all stakeholders is crucial for the success of sustainable construction.

A sustainable construction is beyond the concern of any single discipline. The concept places an emphasis on the integration of all aspects of governance, design in resource consumption, environmental impacts in terms of GHG emissions and pollutions, and well-being in communities, which are essentially associated all social, economic, and environmental aspects.

Meanwhile, for a development to be sustainable, consideration should be given to the full life cycle, that is, from planning, design, construction, operation, maintenance, renovation to demolition.

There are differences between good and sustainable construction. Indicators of good construction practices are normally indications to:

- Meet the users' needs and the owner's requirements;
- Comply with appropriate regulations and standards;
- Achieve good service performance; and

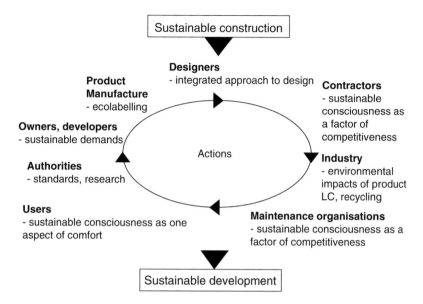

Fig. 1.4 Stakeholders involved in sustainable construction. (Sjostrom & Bakens, 1999)

- Be compatible with the surrounding characteristics.

However, a sustainable construction should try to balance the needs of a good practice as shown above with consideration to:

- Include sustainability principles and climate change into governance for construction development;
- Have a long-term and economically viable and sustainable outcome from a construction development;
- Minimize consumption of resources;
- Minimize atmospheric emissions as well as pollution and wastes;
- Enhance or at least maintain biodiversity of surroundings; and
- Maintain the ethics and value of people, culture, and heritages.

In more detail, governance includes the establishment and oversight of institutions in relation to organizational structures and management represented by policies, guidance, processes, and decision-makings that take into account sustainability and climate change. It increasingly becomes important to take actions in governance to consider both climate mitigation and adaptation.

Sustainable construction should be economically viable and value-generating in a long run considering the longevity of buildings and infrastructure involved with ongoing life cycle asset management. It should minimize the debt that may affect the ability to meet the needs of next generations.

Sustainable construction should minimize the consumption of resources by effectively improving the efficiency, renewability, and recyclability of resources uses. For buildings as an example, efficiency can be improved by enhancing:

Functionality: The issues are floor area per occupant, ceiling height, and adaptability to information technology equipment;

Durability: The issues are the life of the structural and nonstructural element, exterior and interior finishes, equipment, and the reliability of the emergency systems;

Flexibility: The issues are consideration for future renovation and changes of usage; key parameters are floor-to-floor height and adaptability of the floor layout;

Renewability: It is represented by maximising the energy sourced from renewable resources, to reduce the consumption of primary resources. This may further enhanced by increasing the energy consumption efficiency in construction and operation processes, and reduce embodied energy in material selections; and

Recyclability: It is represented by the water reuse and recycling. While it reduces the impact on environment, it also extends the capacity to resist the impact of climate change in terms of increasing scarcity of freshwater resources. The ability to harvest water will further enhance the capacity.

Recyclability is also represented by the reuse and recycle of materials, for example, from construction material wastes. A number of factors need to be considered, such as:

- Recycled and reused: materials with postconsumer and postindustrial contents or from disposal wastes;
- Natural and renewable: materials harvested from sustainably managed sources, such as timber plantation;
- Resource-efficient: materials extracted and manufactured with minimized energy consumption and CO_2/waste generation;
- Locally available: less transportation of materials;
- Durable: product/material has a long service life;
- Amount of materials recyclable at the end of service life;
- Amount of packaging and installation waste; and
- Disposal problems at the end of its service life.

Sustainable construction should minimize carbon emission and pollution discharged to the environment (atmosphere, water bodies, and lands), with the intent to aim zero-emission construction development, by optimal design, treatment, and reuse or recycle. Proper management of carbon emission, water, land, and wastes should be in place.

Sustainable construction should also take into account maintenance or enhancement of biodiversity, which would not be lost as a result of the use of resources, in particular, the use of land. Construction should maintain or enhance the functioning of ecosystems, for example, existing watercourse, natural reserves, and estuaries.

Sustainable construction ensures not only the safety, health, and well-being of workforces by educations and changing behaviors, but also living quality and prosperity of local communities. Construction development should meet both current and future needs of communities without compromising the amenity of other communities. Meanwhile, local culture and heritage should also be respected in sustainable construction.

Most sustainable development indicators in building construction can be categorized into the following groups:

Resource consumption:

- Energy: Greater efficiency in production, construction, and operation is a key requirement for sustainable construction. It is also the source critical to the reduction of GHG emission.
- Materials: Construction is a material-intensive process, consuming a large quantity of mineral resources, which may create a direct impact on biodiversity that is one of the crucial drivers to maintain the life support function of nature. The process and transport of materials also consume energy, creating more embodied energy. The reduction in the use of natural resources for construction materials and the reduction in embodied energy are important for sustainable construction, also in the sense of the conservation of the life support function.
- Water: Lack of water resource has created a challenge on the aspects of the liveability and quality of life. Improvement in water saving and efficiency through water harvest and recycle is an effective step in sustainable construction.
- Land: Land resources are the basis for (human) living systems and provide soil, energy, water, and the opportunity for all human activity. Human settlement and urbanization have been constantly consuming limited land resource, creating a direct impact on the life support function of nature. Sustainable land use for construction may include efficient use of land, design for long service life, and adaptation or conversion of existing buildings. It should be mentioned that the land use may also be interacted with social issues.

Products and buildings:

- Indoor environmental quality: air quality, ventilation, thermal comfort, lighting, and acoustics;
- Service quality of the buildings: adaptability, reliability, controllability, and serviceability; and
- Process and management.

Environmental measures of building construction were also described quantitatively in another term in the Comprehensive Assessment System for Building Environmental Efficiency (Chen et al., 2006) developed by Japan Sustainable Building Consortium. The building environment is divided into two spaces:

- Inside the site boundary: The building construction effort is to improve the environment for the users. This is labeled as Q = "Building Environmental Quality and Performance."

- Outside the site boundary: The building construction is impacting on the rest of the environment. This is labeled as L = "Building Environmental Loadings".

"Building Environmental Quality and Performance" is divided into three categories:

- Indoor environment: It includes acoustics, thermal comfort, lighting, and air quality and ensures comfortable, healthy, and safe indoor environment.
- Quality of service: It includes serviceability, reliability, and adaptability and ensures long service life.
- On-site outdoor environment: It includes preservation, landscape, and amenity and ensures creating a richer townscape and ecosystem.

"Building Environmental Loadings" is also divided into three categories:

- Energy: It includes thermal load, utilization, efficiency, and operation and ensures conserving energy and water.
- Resources: It includes materials and wastes and ensures using resources sparingly and reducing waste.
- Off-site environment: It includes air pollution, noise, light obstruction, light pollution, heat island effects, and loads on local infrastructure and ensures consideration of the global, local, and surrounding environment.

For example, Fig. 1.5 uses a radar chart to describe environmental quality and performance and environmental loading.

The sustainable construction is able to provide environmental, economic, and health and community benefits (Papadopoulos & Giama, 2009), which can be summarized as given below:

Environmental benefits:

- Enhance and conserve ecosystems and biodiversity;
- Improve air and water quality;
- Reduce solid waste;
- Conserve natural resources; and
- Optimize environmental effect on buildings over their life span, "from cradle to grave."

Economic benefits:

- Reduce operational costs;
- Improve asset value and increase profits;
- Improve employees' productivity and inhabitants' satisfaction; and
- Optimize life cycle economic performance.

Health and community benefits:

- Improve air quality and thermal and acoustic environment;
- Improve occupant comfort and health;

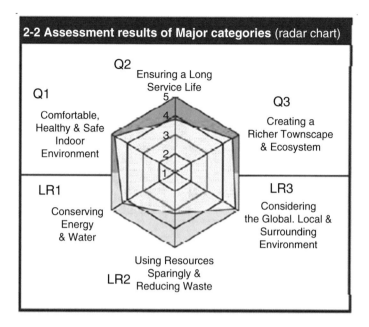

Fig. 1.5 An example of environmental quality and loading. (IBEC, 2007)

- Minimize pressures on the local community infrastructure; and
- Achieve better quality of life.

1.4.2 Life Cycle Approach in Sustainable Construction

Life cycle assessment (LCA) is a methodology to evaluate performance of any level of systems over its service life. It is also called "cradle-to-grave cycle" or "cradle-to-cradle cycle approach" that considers recycle of building materials or products, such as those shown in Fig. 1.6, which describes all the elements of a building life cycle. It is an important approach in evaluating sustainable development and construction and assesses the total impacts or environmental load of a product over its entire life.

ISO 14040, an international standard on LCA, defined four steps that include:

- Define the goal and scope: purposes, audiences, and system boundary;
- Create an life cycle inventory (LCI): collecting data for each product and process in association with relevant inputs and outputs of energy and mass flow (e.g., building materials), emission to air or pollutions to land and water bodies (known as environmental loads);
- Assess the life cycle impact: evaluation of potential environmental impacts and consumed resources in the defined systems, which covers:

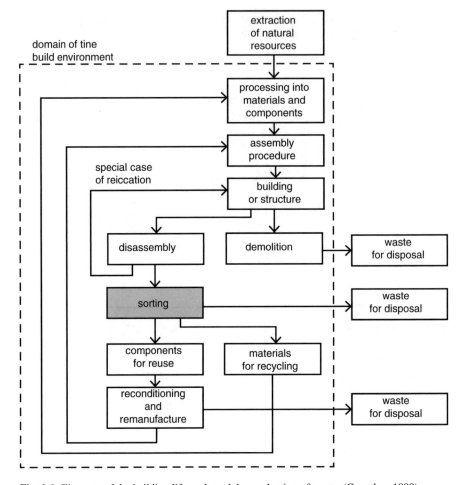

Fig. 1.6 Elements of the building life cycle and the production of waste. (Crowther, 1999)

- – Selection of impact categories,
- – Assignment or classification of LCI results—CO_2, waste etc.,
- – Modeling of category indicators: aggregation into category indicators by normalization and weighting; and

• Interpret the results: identification of significant issues, evaluation of findings, and recommendations.

The LCA in construction can be divided into two areas, that is, building material and component combinations and the whole process of construction (Ortiz et al., 2009).

In general, LCA can be done for a number of different performance indicators, such as monetary cost, energy, ecological footprint, carbon emission, etc., as shown in Fig. 1.7. For example, the manufacturing of a product will require the raw material (embodied with energy, material, and waste), transport (embodied with energy

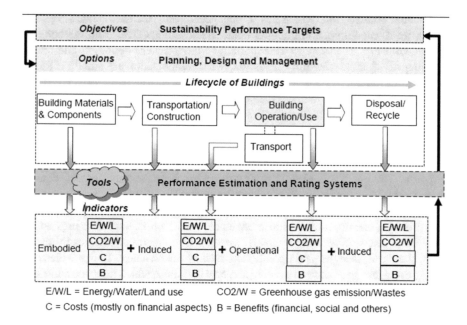

E/W/L = Energy/Water/Land use   CO2/W = Greenhouse gas emission/Wastes
C = Costs (mostly on financial aspects) B = Benefits (financial, social and others)

Fig. 1.7 Life cycle performance–based building sustainability assessment

and pollutants), and processing (embodied with energy and waste). The product installation and use may require more energy, and at the end of its life, there are problems of recycling and/or disposal.

LCA often focuses on one of the indicators described above. Morrissey and Horne (2011) investigated designs of energy-efficient residential buildings, only to minimize the total construction and operating costs over the service life of buildings with consideration of benefit due to energy saving as a result of increased energy efficiency. Meanwhile, assessment can be focused on the energy consumption and associated CO_2 in material production and construction processes, or "cradle-to-site" cycle assessment (Monahan & Powell, 2011), or a broader impact, related to not only energy consumption, but also resource consumption and pollutant discharge (Liu et al., 2010). It should be mentioned, although the above description is more relevant to environmental issues, similar approaches could be applied to deal with social and economic issues.

1.4.3   Challenges of Sustainable Construction

Sustainable construction is developed in various ways, in accordance with regional primary environmental problems, implementation strategy, and nature and state of new and existing buildings. Management and organization is a key aspect of sustainable construction, which has to engage not only technical issues, but also social, legal, economic, and political matters. Difficulties arise from the complex interrelationship. Therefore, it is important to have a clear definition and understanding of what sustainability is sought, what relationships are among different factors contributing to sustainability, what are the indicators and measures, etc. In addition, sustainable development management framework and decision-making protocol have to be established.

A framework to assess the performance of materials, products, and buildings is crucial for sustainable construction. Issues may arise from a wide variety of factors, with some related to physical properties and other related to human behavior and response. The breath of relevant performance has been, to some extent, demonstrated in building environment efficiency as shown in Fig. 1.8.

Emerging technology, especially new energy saving and renewable technology, may provide great opportunities for sustainable construction, such as energy autonomous and self-sufficient buildings. However, it also creates challenges in regulations, standards, designs, and operation as well as maintenance. It also widens the energy efficiency gap between existing and new buildings, which has to be addressed by retrofitting.

Selection of materials—raw, renewable, or recycled—based on environmental performance, including embodied energy and carbon footprints, in addition to cost/benefits and building safety over service life cycle, becomes one of the core contents in sustainable construction.

Resource-consciousness and efficiency on the aspect of water and land use have to be considered in sustainable construction.

Integration of transport infrastructure into sustainable construction will become important issue with urbanization and growing population.

Social and cultural issues become a part of sustainable construction. Sustainable community is closely interconnected with sustainable construction.

Climate change may have a significant impact on the current practices of sustainable construction. Climate-adapted sustainable construction may be required.

1.5 Summary

This chapter provided an introduction to the sustainability and sustainable development as a broader perspective followed by the specific focus on the sustainable development in construction sector. The TBL approach of sustainability assessment, key indicators, key principles, and themes are also discussed to demonstrate how sustainable development can be measured for different sectors. The sustainable development for engineering design and construction has been given a special attention with the key factors for each bottom line (economic, environmental, and social) in sustainable construction. Finally, a life cycle approach is introduced to assess the sustainable development in construction, which considers various new and emerging technologies in sustainable construction.

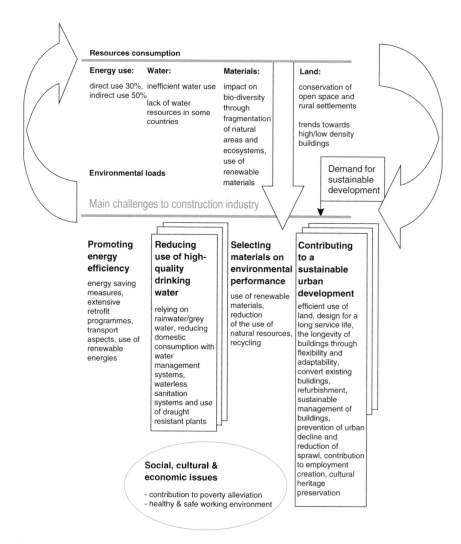

Fig. 1.8 Challenges in sustainable construction. (Sjostrom & Bakens, 1999)

1.6   Discussion Queries

It was requested the readers to understand the key elements in the definition of sustainability in terms of economic, social, and ecological terms. The answers for these queries are as follows:

Economic term: Sustainability is defined as the way that human manages their **economy** to **preserve** its **productiveness and prosperity** through efficiency, investment, diversification, and **balance of** internal demand and external supply **now as well as in the future.**

Ecological term: Sustainability is defined as the way that **human interacts** with the **biosphere** to maintain its **life support function** on the aspect of biological diversity, ecosystem conservation, and regional interconnectedness.

Social term: Sustainability is defined as the way that may improve the **social capital through the encouragement** of **social justices, education, equality, and participative democracy**.

	Economic term	Ecological term	Social term
System	Economy	Human	Social capital
State	Productiveness and prosperity	Life support function	Social justices, education, equality, and participative democracy
Process	???	Interacts	Encouragement
Target	Preserve, balance	Biological diversity, ecosystem conservation, and regional interconnectedness	Improve
Time	Now as well as in the future	???	???

References

Ashe, B., Newton, P. W., Enker, R., Bell, J., Apelt, R., Hough, R., . . . Davis, M. (2003). Sustainability and the building code of Australia.

Board, A. B. C. (2009). *Australian Building Codes Board*. Retrieved February/08.

Chen, Z., Clements-Croome, D., Hong, J., Li, H., & Xu, Q. (2006). A multicriteria lifespan energy efficiency approach to intelligent building assessment. *Energy and Buildings, 38*(5), 393–409.

CIB & UNEP-TETC (2002). "Agenda 21 for Sustainable Construction for Developing Countries –A Discussion Document."International Council for Research and Innovation in Building and Construction (CIB) andUNEP –IETC, Boutek Report No.Bou/E0204

COAG (1997). *Principles and Guidelines for National Standard Setting and Regulatory Action by Ministerial Councils and StandardSetting Bodies*. Council of Australian Governments. Canberra, Australia.

Crowther, P. (1999). *Design for disassembly: An architectural strategy*. Queensland University of Technology.

Dabirian, S., Khanzadi, M., & Taheriattar, R. (2017). Qualitative modeling of sustainability performance in construction projects considering productivity approach. *International Journal of Civil Engineering, 15*(8), 1143–1158.

Dresner, S. (2008). *The principles of sustainability*. Earthscan.

Du Plessis, C. (2002). *Agenda 21 for sustainable construction in developing countries*. CSIR Report BOU E, 204, pp. 2–5.

Estes, R. J. (2010). Toward sustainable development: From theory to praxis. In *Transnational social work practice* (p. 76). Columbia University Press.

Fricker, A. (1998). Measuring up to sustainability. *Futures, 30*(4), 367–375.

IBEC. (2007). *CASBEE for home (detached house) – Technical manual*. Retrieved from.

Industry, A., & Committee, S. (2020). National industry insights report: 2018/19 national overview.

Kibert, C. J. (1994). *Establishing principles and model for sustainable construction*. Paper presented at the First International Conference on Sustainable Construction, Tampa.

Littig, B., & Griessler, E. (2005). Social sustainability: A catchword between political pragmatism and social theory. *International Journal of Sustainable Development, 8*(1–2), 65–79.

Liu, M., Li, B., & Yao, R. (2010). A generic model of exergy assessment for the environmental impact of building lifecycle. *Energy and Buildings, 42*(9), 1482–1490.

Lumley, S., & Armstrong, P. (2004). Some of the nineteenth century origins of the sustainability concept. *Environment, Development and Sustainability, 6*(3), 367–378.

Monahan, J., & Powell, J. C. (2011). An embodied carbon and energy analysis of modern methods of construction in housing: A case study using a lifecycle assessment framework. *Energy and Buildings, 43*(1), 179–188.

Morrissey, J., & Horne, R. E. (2011). Life cycle cost implications of energy efficiency measures in new residential buildings. *Energy and Buildings, 43*(4), 915–924.

Ortiz, O., Castells, F., & Sonnemann, G. (2009). Sustainability in the construction industry: A review of recent developments based on LCA. *Construction and Building Materials, 23*(1), 28–39.

Pachauri, R. K., Allen, M. R., Barros, V. R., Broome, J., Cramer, W., Christ, R., Church, J. A., Clarke, L., Dahe, Q., Dasgupta, P., Dubash, N. K., Edenhofer, O., Elgizouli, I., Field, C. B., Forster, P., Friedlingstein, P., Fuglestvedt, J., Gomez-Echeverri, L., Hallegatte, S., ... van Ypserle, J.-P. (2014). *Climate change 2014: Synthesis report. Contribution of Working Groups I, II and III to the fifth assessment report of the Intergovernmental Panel on Climate Change*. IPCC.

Papadopoulos, A., & Giama, E. (2009). Rating systems for counting buildings' environmental performance. *International Journal of Sustainable Energy, 28*(1–3), 29–43.

Parkin, J., & Sharma, D. (1999). *Infrastructure planning*. Thomas Telford.

Plank, R. (2008). The principles of sustainable construction. *The IES Journal Part A: Civil & Structural Engineering, 1*(4), 301–307.

Sjostrom, C., & Bakens, W. (1999). CIB Agenda 21 for sustainable construction: Why, how and what. *Building Research & Information, 27*(6), 347–353.

Team, E., & BRE, A. G. (1999). A comparison of visions from various countries.

Waas, T., Hugé, J., Block, T., Wright, T., Benitez-Capistros, F., & Verbruggen, A. (2014). Sustainability assessment and indicators: Tools in a decision-making strategy for sustainable development. *Sustainability, 6*(9), 5512–5534.

WCED, S. W. S. (1987). World commission on environment and development. In *Our common future* (Vol. 17, pp. 1–91). Oxford University Press.

Chapter 2
International and National Sustainable Developments

2.1 Introduction

The global awareness of sustainability and sustainable development has emerged in the past four decades or so with all sectors of our society developing and interpreting the principles of sustainability as per their specific context. This topic outlines some of the key international developments by organisations such as the United Nations (UN), the International Standard Organisation (ISO), the Organisation for Economic Co-operation and Development (OECD) as well as the national developments contributed by federal, state and local governments in Australia.

2.2 International Developments

2.2.1 History of Sustainable Development

In 1972, the UN Conference on the Human Environment in Stockholm (the "Stockholm Conference") was held to address the global environmental problems, including the concerns of transboundary pollution, such as acid rain. The main purpose of the conference was to promote and advance the guidelines for taking actions by governments and international organizations to remedy and prevent impairment of the environment. It was reported that the Stockholm conference was a great success with the attendance of 130 countries, and the conference generated three major products, which are the action plan, the United Nations Environment Programme (UNEP) and the Stockholm Declaration (Sullivan, 1972).

The action plan identified the environmental issues that require international attention and launched a global environmental assessment programme, known as **Earthwatch**. The priority recommendation of the action plan was the developments

© Springer Nature Switzerland AG 2021
X. Wang, S. Ramakrishnan, *Environmental Sustainability in Building Design and Construction*, https://doi.org/10.1007/978-3-030-76231-5_2

of subsequent international environmental agreements and shaping the agenda for the UN environmental programme. It was reported that the Stockholm Declaration helped lay the foundation for sustainable development with the proposal of 26 principles. The principles mainly address the importance of protecting and improving the environment for present and future generations (known as principle 1), safeguarding natural resources and wild habitat (principles 2,4) and prevention of pollution and discharge of toxic substances (principles 6,7). While the three major products reported above had an implication for global justice, the most significant impact of the conference was raising awareness on the potential degree of human influence on the global environment, including the climate.

In 1987, the World Commission on Environment and Development (WCED) was established as a special commission to publish a report on the environment and global issues faced by 2000 and beyond with the proposed strategies for sustainable development. The published report, entitled "Our Common Future" (also known as the Brundtland Report), brought the concept of sustainable development by defining it as "development which meets the needs of the present without compromising the ability of future generations to meet their own needs" (Keeble, 1988). The Brundland report was considered as the basic material in the preparation for the adoption of environmental perspective to the year 2000 and beyond. The report emphasised the inseparable relationship between environmental problems and socioeconomic issues, which lead to the call for a UN Conference on Environment and Development.

In 1992, the Earth Summit, also known as the UN Conference on Environment and Development, was held in Rio de Janeiro with the participation of 180 countries (Grubb, Koch, Thomson, Sullivan, & Munson, 2019). The main purpose of the conference was to produce a broad agenda to address some of the most pressing problems facing the world by 2000 and beyond and to discuss how to achieve sustainable development. The summit concluded that the proposed concept of sustainable development can be an attainable goal by the global population, regardless of whether they are on a local, national or international level. It also emphasised that an integrated and balanced approach between the economic, social and environmental pillars is required for a sustained human life on the planet. The Earth Summit had many achievements, including five major setouts of Rio declaration on environment and development, the UN Framework Convention on Climate Change (UNFCCC), Agenda 21, the statement of forest principles and the convention on biological diversity; Agenda 21 was a major result of the conference proposing a programme of action calling for new strategies to achieve overall sustainable development by the twenty-first century.

The UNFCCC proposed its ultimate objective as stabilizing the atmospheric greenhouse gas concentrations "at a level that would prevent dangerous anthropogenic [human] interference with the climate system." Given the different countries contribute to climate change at different levels and their capability of reducing the greenhouse gas (GHG) emissions may vary, the governments agreed for a "common but differentiated responsibilities." With this variability, the developed countries agreed to "take the lead," followed by the developing countries, with the assistance of developed countries, in combating climate change. With this agreement, the developed countries made a non-binding "aim" of reducing their emissions to 1990 levels by 2000. After the Earth Summit, the status of each country on reducing the GHG levels is monitored by the UN Commission on Sustainable Development set by the United Nations.

In 1996, a special session of the UN General Assembly (Earth Summit +5) was held to review and appraise the implementation of Agenda 21 and assess how well countries and international organizations have responded to the challenges of the Earth Summit (Grubb et al., 2019).

In 1996, followed by the Earth Summit, the second conference on sustainable human settlements (HABITAT II) was held in Istanbul. The conference has led to the development of an action plan for matters related to the construction industry and how governments should encourage this industry to apply sustainable developments. The action plan focused on the promotion of sustainable building materials and its production along with the implementation of sustainable development at all phases of construction, such as planning, design, construction, maintenance and rehabilitation and procurement (Okpala, 1996).

In 1997, the UNFCCC adopts the Kyoto Protocol, which set binding targets to reduce GHG emissions to an average of 5% to the 1990 levels by 2008–2012, primarily for 37 industrialized countries and the European community. These 37 industrialized countries with binding targets account for 60% of developed country emissions and about a quarter of global emissions (Protocol, 1997a, b). The key provisions were provided with the flexibility to meet the targets by using either of the following three market-based mechanisms:

1. International emission trading: Trading of emission allowances between countries
2. Joint implementation and clean development mechanism: Crediting emission reductions from projects that involve joint implementation between countries and focused on clean energy development mechanisms
3. Other flexibility provisions: Providing emission targets as five years average than single year limits; including all six greenhouse gases rather than carbon dioxide for counting purposes; crediting carbon sequestration opportunities such as forests and farmlands

In 2000, the United Nations Millennium Summit supported the principles of sustainable development, including those set out in Agenda 21 were reaffirmed (Summit, 2000).

In 2002, the Rio+10, also known as the Johannesburg Summit, took place in South Africa during August–September 2002. Secretary-General Kofi Annan identified five themes as critical areas for long-term development, involving different sectors, organizations and disciplines, as well as the interactions between economic, environmental and social factors (Von Frantzius, 2004). These themes are water, energy, health, agriculture and biodiversity. Commitment to sustainable development was reaffirmed again in the Johannesburg Declaration. The Johannesburg Plan of Implementation promoted the integration of the three pillars of sustainable development: economic development, social development and environmental protection.

It indicated that the objectives of this summit were poverty eradication, changing unsustainable patterns of production and consumption and protecting and managing natural resources, which are also essential requirements for sustainable development (Hens & Nath, 2003).

In 2005, a follow-up summit meeting to Millennium summit was held by the UN General Assembly. This summit reiterated that "sustainable development is a key element of the overarching framework for United Nations activities, in particular for achieving the internationally agreed development goals"(Bellamy, 2006; Wheeler, 2005).

In 2012, the UN organised the Conference on Sustainable Development in Rio de Janeiro, known as Rio+20, with themes more focused on building a green economy in the context of sustainable development and poverty eradication and an institutional framework for sustainable development (Bulkeley et al., 2013; Clémençon, 2012). The Future We Want: a High-level Political Forum on Sustainable Development to subsequently replace the Commission on Sustainable Development (CSD) (1993) and launch a process to develop a set of Sustainable Development Goals (SDGs).

In 2015, the UN General Assembly formally adopted a set of 17 Sustainable Development Goals (SDGs) and 169 associated targets with the integrated and transformative 2030 Agenda for Sustainable Development (GA, 2015).

2.2.2   UN Policy on Sustainable Development

Agenda 21 is a comprehensive plan of action proposed at the United Nations Conference on Environment and Development (UNCED) held in Rio de Janeiro, Brazil, 1992. These actions are taken globally, nationally and locally by more than 178 governments (Sitarz, 1993). Agenda 21 has many items that impact on the construction sector and the items relevant to the development of Building Codes are (Ashe et al., 2003):

Chapter 4:

- Developing national policies and strategies to encourage changes in unsustainable consumption patterns

Chapter 7:

- Promoting adequate shelter for all
- Promoting human settlements planning and management in disaster-prone areas
- Promoting sustainable construction activities

Chapter 9:

- Promoting sustainable development and protection of the atmosphere through energy development, efficiency and consumption

Chapter 20:

- Promoting prevention and minimisation of hazardous waste

Chapter 21:

- Minimising waste
- Maximising environmentally sound waste reuse and recycling
- Promoting environmentally sound waste disposal and treatment
- Extending the waste service coverage

Chapter 30:

- Promoting cleaner production

Chapter 40:

- Providing information for decision-making (e.g. indicators for sustainable development)

The **Earth Summit+5** is a special session of the UN General Assembly to review and appraise the implementation of Agenda 21 (Lafferty & Eckerberg, 2013).

The Commission on Sustainable Development (CSD) was created to follow up with the UNCED and to monitor and report on the implementation of the agreements at local, national, regional and international levels. The CSD Work Program on Indicators of Sustainable Development (1995–2000) resulted in a description of key sustainable development themes and the development of indicators for use in the decision-making processes at the national level (Spangenberg, 2002). A revised set of CSD indicators was finalized in 2006. The revised edition contains 96 indicators, including a subset of 50 core indicators (de Oliveira & Trindade, 2018).

The revised CSD indicators were considered in the following themes (UNDESA, 2007):

Social

- Poverty
- Governance
- Health
- Education
- Demographics

Environmental

- Natural hazards
- Atmosphere
- Land
- Oceans, seas and coasts
- Freshwater
- Biodiversity

Economic

- Economic development
- Global economic partnership
- Consumption and production patterns.

Indicators that are relevant to building and construction are (Ashe et al., 2003):

- Environmental – Natural Hazard

    - Percentage of population living in hazard-prone areas
    - Human and economic loss due to natural disasters

- Environmental – Atmosphere

    - CO_2 emissions
    - Emissions of greenhouse gases
    - Consumption of ozone-depleting substances
    - Ambient concentration of air pollutants in urban areas

- Environmental – Land

    - Land use change
    - Lland degradation

- Environmental – Freshwater

    - Proportion of total water resources used
    - Presence of faecal coliforms in freshwater

- Economic – Consumption and production patterns – material consumption

    - Material intensity of the economy
    - Domestic material consumption

- Economic – Consumption and production patterns – energy use

    - Annual energy consumption, total and by main user category
    - Intensity of energy use, total and by economic activity,
    - Share of consumption of renewable energy resources

- Economic – consumption and production patterns – waste generation and management

    - Generation of wastes: Solid waste, hazardous waste
    - Waste treatment and disposal

- Economic – consumption and production patterns – transportation

    - Modal split of passenger transportation
    - Modal split of freight transportation
    - Energy intensity of transportation

Other social themes may also be directly or indirectly related to building and construction, such as:

- Social – Poverty

  - Proportion of urban population living in slums
  - Percentage of population using solid fuels for cooking

- Social – Demographics

  - Population growth rate

The **Kyoto protocol**, as mentioned in the history of sustainable development, is an attempt by UNFCCC to mitigate the threat of global warming and to stabilise GHG emissions at 1990 levels by the year of 2000. This agreement was made in December 1997 at the third Conference of Parties (COP3), held in Kyoto, Japan. Industrial nations agreed to reduce their collective emissions of greenhouse gases by 5.2% from 1990 levels by the period 2008–2012. The Kyoto Protocol was endorsed by 160 countries

The protocol was proposed to be legally binding when at least 55 countries sign up for it. This includes the developed countries that are responsible for at least 55% of greenhouse gas emissions from the industrialised world. Here, the six greenhouse gases of carbon dioxide, methane, nitrous oxide, hydrofluorocarbons (HFCs), perfluorocarbons (PFCs) and sulphur hexafluoride were considered. The global emission reductions by 5.2% was to be achieved by differential reductions for individual countries.

The emission reduction criteria varied between different nations, with the European Union (EU) countries, Switzerland and the majority of Central and Eastern European nations agreeing for 8% reductions. The United States agreed to cut their emissions by 7%, followed by Canada, Japan, Poland and Hungary by 6%. Countries like Russia, New Zealand and Ukraine were required to stabilise their emissions. On the other hand, Norway, Australia and Iceland were allowed for a slight increase in their emissions, given at a reduced rate to the emission trends at the agreement time. In this regard, Australia's commitment was to limit the increase in greenhouse gas emission to 8% of its 1990 emission level by the period 2008–2012.

Habitat II was formulated as a second international action plan, followed by Agenda 21, to specifically address the role of human settlement and building construction in sustainable development (Okpala, 1996). This particular agenda sets out:

- Specific actions to be taken by the government and construction industry to adopt sustainable development in planning, design, construction maintenance and rehabilitation phases
- Promotion and use of sustainable building materials
- Production of sustainable materials

2.2.3   CIB Agenda 21 on Sustainable Construction

In 1999, the International Council for Research and Innovation in Building and Construction (CIB) published its Agenda 21 on sustainable construction (Carassus, 2004). It was intended to be an intermediary between Habitat II Agenda and the National Agendas. It discusses (Ashe et al., 2003):

- Concept of sustainable construction
- Issues and challenges of sustainable construction
- Resulting challenges and actions

The main objective of the CIB agenda was to develop a universal framework that would be applied to all national, regional and sub-sectoral agendas as well as to provide with a reliable source for research and development activities related to sustainable construction. The agenda recommends the building codes to provide scalable and measurable tools for assessing environmental performance in building construction. This includes: (Ashe et al., 2003)

- Performance-based standards in the building codes
- "Green" certification and eco-labelling systems based on life cycle analysis

2.2.4   OECD Policies on Sustainable Development

The Organisation for Economic Co-operation and Development (OECD) had been working on "Sustainable Building Project," which is aimed at assisting the member countries by providing guidance for the design of government policies to address three major environment impacts of the building sector. These are CO_2 emissions caused by the building sector, waste generation during the construction and demolition processes and indoor air pollution (Kibert, 2001). The summary on the current environmental policies relevant to three major environmental impacts is as follows (Jörgens, 2012):

- CO_2 emissions: The policy instruments primarily targeted the new buildings for reducing the CO_2 emissions, while the government intervention for upgrading the existing buildings were limited. The information tools related to environmental labelling were increased; however, the economic instructions were limited.
- Waste generation in construction and demolition stages: The policy instruments are implemented to reduce waste generation during the demolition stage. Regulatory instruments such as mandatory separation of waste types, landfill ban and landfill tax were widely issued in the European countries. Few policy instruments were developed to regulate activities at the upstream stages (i.e. before design and construction).

- Indoor pollution: The most commonly used policy instrument for regulating indoor air pollution was to set target values of maximum pollutant concentrations. The environmental labelling schemes covering the issue of indoor air quality applied in several countries. Regulations on the quality of building materials were also implemented in four European countries.

2.2.5 ISO Policies on Sustainable Development

The International Standards Organisation (ISO) has developed two series of standards and one guideline related to sustainability (Ashe et al., 2003).

ISO 15686 Building and Constructed Assets – Service Life Planning: This standard identifies and establishes general principles for service life planning in building construction and proposes a systematic framework for undertaking service life planning of a construction throughout its lifecycle. The following sections of the standard addresses the concept of service life planning:

- ISO 15686-1: 2011, Buildings and constructed assets – Service life planning – Part 1: General principles and framework
- ISO 15686-5: 2017, Buildings and constructed assets – Service life planning – Part 5: Life-cycle costing
- ISO 15686-6: 2004, Buildings and constructed assets – Service life planning – Part 6: Procedures for considering environmental impacts

ISO 14000 Series on Environmental Management Systems: A series of voluntary standards that provide a structure for managing environmental impact.

- ISO 14001:2015, Environmental management systems – Requirements with guidance for use
- ISO 14004: 2016, Environmental management systems – General guidelines on implementation
- ISO 14020: 2000, Environmental labels and declarations — General principles
- ISO 14031: 2013, Environmental management — Environmental performance evaluation — Guidelines
- ISO 14040: 2006, Environmental management — Life cycle assessment — Principles and framework

ISO Guide 64: 1997, Guide for the inclusion of environmental aspects in product standards

2.2.6 Melbourne Principles for Sustainable Cities

The Melbourne Principles for Sustainable Cities was developed to provide a strategic framework for sustainable urban development, provide cities with a foundation for the integration of international, national and local programmes, identify and address gaps, as well as realise synergies through partnerships. It was proposed in Melbourne, Australia during an international Charette in April 2002, sponsored by the United Nations Environment Programme (UNEP) and the International Council for Local Environmental Initiatives. The principles include (UNEP, 2002):

- Provision of a long-term vision for cities based on: sustainability; intergenerational, social, economic and political equity and their individuality.
- Promote sustainable production and consumption through appropriate use of environmentally sound technologies and effective demand management.
- Expand and enable cooperative networks to work towards a common, sustainable future.
- Enable continual improvement based on accountability, transparency and good governance.
- Build on the characteristics of ecosystems in the development and nurturing of healthy and sustainable cities.
- Recognise the intrinsic value of biodiversity and natural ecosystems, and protect and restore them. Enable communities to minimise their ecological footprint.
- Recognise and build on the distinctive characteristics of cities, including their human and cultural values, history and natural systems.
- Achieve long-term economic and social security.
- Empower people and foster participation.

2.3 National Developments

2.3.1 Introduction

Australia has three levels of governments: federal, state and territory and local governments. The constitution assigned the responsibilities for building and construction to the state and territory governments. There are six states and two territory governments in Australia. The states are New South Wales (NSW), Victoria, South Australia, Queensland, Western Australia and Tasmania. The territory governments are Australian Capital Territory (ACT) and Northern Territory (NT). Each state and territory governments assign different parts of this responsibility to local governments. Consequently, with regard to sustainable construction, all three levels of governments have taken various initiatives, sometimes together, some other times independently. This book reviews developments taken by the federal and state and territory governments. Local government activities can only be mentioned as examples, since there are too many local governments for a proper survey to be done (over 600 Australia-wide).

The information presented here may not represent the latest development, since there are continuing national developments at all levels of governments in Australia. More information can be found on various government websites.

2.3.2 *Australia's Commitment to International Sustainable Development*

Australia is a member of several international organisations that promote sustainable development and participates in a number of agreements related to sustainable development. Some of the international organisations are:

- Inter-governmental Panel on Climate Change (IPCC)
- Kyoto Protocol for greenhouse gas emission reductions
- Convention on Biological Diversity and the Biosafety Protocol
- Prior Informed Consent Convention governing trade in hazardous chemicals
- Montreal Protocol on ozone-depleting substances
- Basel Convention on transboundary movement of hazardous wastes

2.3.3 *Council of Australian Government (COAG) Agreements*

In 1992, the Australian government, with endorsement of the Council of Australian Governments (COAG), released its National Strategy for Ecologically Sustainable Development (NSESD) (COAG, 1997). The Council agreed that the future development of all relevant policies and programmes should take place within the framework of the Ecological Sustainable Development (ESD) strategy and the Intergovernmental Agreement on the Environment, which came into effect in May 1992 (Australia, 1997).

Ecological Sustainable Development (ESD) is defined in the Strategy as "using, conserving and enhancing the community's resources so that ecological processes, on which life depends, are maintained, and the total quality of life, now and in the future, can be increased" (Beer, 2003). There are five key ESD principles defined as (Lehmann, 2006):

- Integrating economic and environmental goals in policies and activities
- Ensuring that environmental assets are properly valued
- Providing for equity within and between generations
- Dealing cautiously with risk and irreversibility
- Recognising the global dimension

A range of recommendations were made to guide governments in policy development for ESD in relation to specific industry sectors.

In 1995, Principles and Guidelines for National Standard Setting and Regulatory Action were developed by Ministerial Councils and Standards Setting Bodies.

In 1997, the Council endorsed the Commonwealth's international negotiating position on climate change to achieve an effective environmental outcome. Specific actions were outlined by the prime minister to reduce Australia's greenhouse gas emissions, including (Territory, 1997):

- Reduction of industry emissions, including the expansion of the Greenhouse Challenge Programme, and improving energy codes and standards
- Establishing and further enhancing carbon sinks and in particular the encouragement of plantation establishment
- Reducing transport emissions, including those from private cars
- Encouraging reduction of residential emissions
- Reducing emissions in the Commonwealth's own operations

In 2007, COAG National Reform Agenda was announced. Some relevant agendas are (Greenhouse, 2007):

- *Regulation Reform*: Develop a more harmonised and efficient system of environmental assessment and approval, ensure best practice regulation is applied to the Building Code of Australia (BCA), and remove unnecessary state-based variations to the BCA.
- *Greenhouse Gas Emissions and Energy Reporting*: COAG agreed to establish mandatory national greenhouse gas emissions and energy reporting system, with the detailed design to be settled after the Prime Minister's Task Group on Emissions Trading reports.
- *Climate Change:* COAG endorsed a National Adaptation Framework as the basis for jurisdictional actions on adaptation over the next five to seven years. The framework includes possible actions to assist the most vulnerable sectors and regions, such as agriculture, biodiversity, fisheries, forestry, settlements and infrastructure, coastal, water resources, tourism and health to adapt to the impacts of climate change and actions to establish the Australian Centre for Climate Change Adaptation and CSIRO Adaptation Flagship in order to develop tangible responses to climate change, for example:

  - Identification of how to protect coastal infrastructure from likely changes in storm surge using well-designed sea walls and flood barriers
  - Work towards the design of a heatwave warning system and proposing ways to modify facilities to cater for those most at risk (the elderly)
  - Helping to plan for expanding the use of feedlots by farmers to reduce the exposure of their valuable stock to variation in pasture availability and heat stress
  - Identifying areas in national parks that will provide the best areas for recolonization of plants and animals that have been displaced by climatic changes from their natural locations

In 2008, the COAG developed an agenda known as "Reform Agenda" for reforming and investing for the future, with particular focus on health, water, regulatory reform and the broader productivity agenda. This agenda ensures sustainable water supply and has expanded the CSIRO assessments of Sustainable Yields in order to have a comprehensive scientific assessment of sustainable water yield in all major river systems across the country

The COAG has embraced a new national approach to address climate change through the introduction of national emissions trading scheme (ETS) and complementary policies that achieve emission reductions at least cost. The introduction of the national ETS to achieve emission reductions will constitute the most significant economic and structural reform, since the trade liberalisation and financial market reforms of the 1980s.

The COAG also called for bringing different approaches together on renewable energy targets to combine into one national scheme in order to provide consistency for investors looking to support Australia's renewable energy industry. In addition, the COAG agreed to consider options for a harmonised approach to renewable energy "feed in tariffs" in October 2008.

2.3.4   Federal Australian Government Actions

In 1998, the **National Greenhouse Strategy** was developed to ensure "Australia will actively contribute to the global effort to stabilise greenhouse gas concentrations in the atmosphere at a level that would prevent dangerous interference with the climate system and within a time frame sufficient to allow ecosystems to adapt naturally to climate change, to ensure that food production is not threatened, and enable economic development to proceed in a sustainable way" (Australia, 1998).

The **Environmental Protection and Biodiversity Conservation Act** was proposed in 1998 and passed in 1999. This act requires developments with impacts on special environments, such as land development, are to be referred to the Environment Minister for consideration.

The 516A of the Environment Protection and Biodiversity Conservation Act, 1999 specifies that all annual reports produced by the Commonwealth departments, Parliamentary departments, Commonwealth authorities, Commonwealth companies and other Commonwealth agencies must include a report on environmental matters. The annual reports must (Australian Government, 1999):

- Report how the agency's activities have accorded with the principles of Ecologically Sustainable Development (ESD)
- Identify how their departmental outcomes contributed to ESD
- Document the agency's impacts upon the environment and measures taken to minimise those impacts
- Identify the review mechanisms in place to review and increase the measures the agency takes to minimise its impact upon the environment

The minister must instruct the Department of the Environment, Water, Heritage and the Arts to prepare a "State of the Environment report for Australia" at every five years' interval. The minister must table this report in the parliament. In fact, state of the environment (SoE) reporting occurs at both the national and state/territory level.

National SoE reports collect information regarding the environmental and heritage conditions, trends and pressures for the coverage of the Australian continent, the surrounding seas and Australia's external territories. These reports are based on data and information gathered and interpreted against environmental indicators. The environmental indicators have been grouped into environmental themes. These environmental themes are known as the SoE reporting themes, which are as follows:

- Atmosphere: Covers climate variability and change, stratospheric ozone and urban and rural air quality
- Biodiversity: Covers utilisation and value of biodiversity, pressures on biodiversity (including clearing, changed fire regimes, total grazing pressure and weeds and feral animals), condition of species, habitats and ecological communities and landscapes and their protection and management
- Coasts and oceans: Covers condition of habitats, coastal urban development and coastal water quality, fisheries and aquaculture, introduced marine species and responses to pressures
- Human settlements: Covers population changes, urban development and design, material and energy use, urban water use, transport and the condition of human settlements
- Inland waters: Covers water availability and use of surface water and groundwater, water quality, pressures on aquatic ecosystems (including aquatic biodiversity, river and wetland salinity, introduced species), response of biota to pressures and investments in inland waters
- Land: Covers vegetation, land condition (including soil loss and dust storms, soil carbon, salinity and soil acidity), institutional pressures and responses to pressures
- Natural and cultural heritage: Covers knowledge of heritage, physical condition and integrity of heritage, responses to identify and protect heritage, expertise and skills for managing heritage and community awareness of heritage
- Australian Antarctic Territory: Covers climate, atmosphere and the ice, marine ecosystems, human pressures and Antarctic heritage (including condition of Antarctic heritage sites, risks to Antarctic heritage sites, resources to manage Antarctic heritage sites and Antarctic heritage collections in Australia)

Environment Australia (the Department of Environment and Heritage Protection) was also established **in 1999**. The Environment Australia administers environmental laws, including the Environmental Protection and Biodiversity Conservation Act (EPBC) of 1999. The portfolio also includes the Australian Greenhouse Office and manages a range of voluntary programmes related to building construction (Ashe et al., 2003).

In December 2001, NABERS (the National Australian Buildings Environmental Rating System) was released by Environment Australia as an environmental rating system for residential and commercial buildings. This system is expected to perform a rating system than a design tool, and its scope covers: land, energy, materials, water, waste, interior, resources and transport. Under each category, the scores are provided as per the performance of the building to determine a star rating between one and five. Formerly Australian Greenhouse Office (AGO), and now the Department of Climate Change and Energy Efficiency is the lead Commonwealth agency on greenhouse matters. The agency focuses on the following building construction–related issues (Ashe et al., 2003):

- Energy efficiency in buildings
- Supporting the use of solar power in residential and community buildings

Environment Australia (EA) also sponsors **State of Environment Reports,** which contain data related to Australia's State of the Environment. The first report was issued in 1996 and the second in 2001 (Newton, 2011). These reports provide a once-in-five-year summary of the key issues related to urban Australia and their environmental impacts. Some of the **voluntary programmes** managed by EA include:

- NABERS project to develop a national building rating system
- WasteWise Construction Program
- PATHE/GreenSmart Program (with HIA)
- Recycled Concrete Guide (with CSIRO)
- Awards (with MBA)

Engineers Australia, formerly known as the Institution of Engineers, Australia, published "Sustainable Energy Building and Construction Taskforce Report" (Jefferson, 2006), which explored the ways the building and construction sector might contribute to more sustainable energy practices. It produced a series of recommendations to governments and individual engineers.

The Building Code of Australia (BCA) was a set of technical rules for the design and construction of buildings. Although it provided guidelines on durability and energy efficiency, as given below, it did not address the issue of sustainability and had no reference, recommendations or restrictions on any issue related to sustainability (Ashe et al., 2003).

- **Durability**: Durability was not directly addressed in the BCA, but durability requirements are included in referenced documents to satisfy the primary goals of health, safety and amenity.
- **Energy Efficiency**: Two states and a territory (the Australian Capital Territory, South Australia and Victoria) had measures in their BCA Appendices. A new energy efficiency measure regulatory proposal for housing – 'Energy Efficiency Measures – BCA96 Vol.2 (Housing Provisions)' – was developed and considered by ABCB in 2002.

In 2007, the Australian Building Codes Board included sustainability as a goal for the BCA alongside the existing BCA goals of health, safety and amenity. The BCA2008 introduced provisions for all buildings. In 2007, the Government of Australia also decided to ratify the **Kyoto Protocol**.

In 2008, the Garnaut Review or Garnaut Climate Change Review was commissioned by the Australian Government to evaluate the impacts of climate change on the Australian economy and to recommend medium- to long-term policies and policy frameworks to mitigate and adapt the impact and improve the prospects of sustainable prosperity. It draws the answers on what extent of global mitigation, with Australia playing its proportionate part, provides the greatest excess of gains from reduced risks of climate change over costs of mitigation. A draft report was released in July 2008, applying to the evaluation of the costs and benefits of climate change mitigation, application of the science of climate change to Australia, the international context of Australian mitigation and the Australian mitigation policy. Its final report was completed by the end of 2008.

The Emission Trading Scheme (ETS) was released in March 2008 putting forward a set of principles and design features for an efficient and effective emission trading. This was followed by an Australian Government Green Paper on "Carbon Pollution Reduction Scheme" in July 2008.

2.4   State and Territory Developments

The sustainability and building code of Australia (Ashe et al., 2003) summarises the state and territory developments regarding the sustainability trends and provisions, particularly related to building and construction. The developments in each state and territory with the specific timeframes are provided below:

The **New South Wales** has initiated Ecological Sustainable Development (ESD) under the legislations including the Local Government Act (1993), Environment Planning and Assessment Act (1979) and Protection for Environment Operations Act (1997).

The Sustainable Energy Development Authority was set up in 1996 to reduce greenhouse gas emissions and to develop renewable energy and cogeneration. It was later merged with the Ministry of Energy into the Department of Energy, Utilities and Sustainability in 2004.

Resources NSW was established by the Waste Avoidance and Resource Recovery Act of 2001 to manage avoidance, resource recovery and disposal of waste. The Sustainability Unit of Planning NSW developed a Sustainability Building Index/ Rating Tool (BASIX) intended to be taken up by local governments in 2003.

In 2003, the Greenhouse Gas Reduction Scheme (GGAS) was introduced, a mandatory greenhouse gas emissions trading scheme that aimed to reduce greenhouse gas emissions associated with the production and use of electricity.

In 2005, the Greenhouse Action Plan was released, which included a wide range of government initiatives to address greenhouse emissions caused by many sectors, including energy generation, energy efficiency, agriculture, natural resources, buildings, transport, industrial processes, waste, fugitive emissions, land management and government leadership and energy management policy.

In 2006, the NSW Renewable Energy Target (NRET) was introduced. The scheme required to use renewable energy for a proportion of electricity consumed by NSW consumers. The target renewable energy proportion level was 10% by 2010 and 15% by 2020 of end-use consumption.

Victoria, following the Victorian Greenhouse Strategy, Victoria has mainly focused on climate change response by enhancing building energy efficiency in residential and commercial buildings. The government announced a five star energy rating scheme for residential buildings. Another issue is the management of solid waste management by construction and demolition activities, which were managed by EPA, local councils and EcoRecycle.

Water management was also an issue, and the state focused at the policy level through Victorian Water Strategy to produce a series of strategic direction reports.

In 2001, the Victorian Government made a high-level policy commitment to sustainability, expressed in the publication of "Growing Victoria Together".

In 2005, Victoria's Environmental Sustainability Framework was launched. It outlined the key environmental challenges and identified the strategic directions to become environmentally sustainable.

In South Australia, the Environment Protection Act 1993 provided environmental policies, guidelines and codes of practice for a wide range of environmental issues. The issues related to the building and construction industry include recycling of building and demolition waste, waste on building sites, noise, stormwater pollution, etc. For long-term management of the waste generated in the city, an office of sustainability was established. It later became the Sustainability and Climate Change Division of the Department of Premier and Cabinet.

The Development Act 1993 had developed specific objectives for building rules as follows:

'To encourage the management of the natural and constructed environment in an ecologically sustainable manner'. In addition, 'To facilitate sustainable development and the protection of the environment'. Under this act, significant packages were provided for stormwater management and wind farms projects.

The state had a significant commitment to reduce greenhouse gas emissions and improve the energy efficiency of buildings. Energy efficiency requirements for buildings were introduced on 1 January 2003.

In 2007, Tracking Climate Change: South Australia's Greenhouse Strategy 2007–2020 was released. The strategy was to be implemented through three avenues, which include reducing greenhouse emission, adapting to climate change, and innovating in markets, technologies, institutions and the way we live. Infrastructure, urban planning, built environment and natural resources were among the sectors facing challenges in adapting to climate changes. Especially for buildings, it required to:

- Develop high performance green standards for building design, construction and operation
- Optimize the energy performance and subsequent cost-effectiveness of buildings
- Increase market and community awareness of the benefits of improved building performance
- Develop sustainable built environments that are responsive to climate change

In Climate Change and Greenhouse Emissions Reduction Act 2007, three targets were set, which include GHG emission reduction in the state by at least 60% of 1990 levels by the end of 2050, renewable energy generated by at least 20% of electricity generation, and renewable energy consumed by at least 20% of electricity consumption.

In Queensland, the Environmental Protection Act 1994 provided policies for environmental concerns such as air, noise, water, and waste management. These policies subsequently affected the building construction sector.

In 1997, the Integrated Planning Act 1997 was established as "a framework to integrate planning and development assessment so that development and its effects are managed in a way that is ecologically sustainable, and for related purposes". In this framework, "Development" is defined as "carrying out building work." "Ecological sustainability' is defined as a "balance that integrates:

- Protection of ecological processes and natural systems at local, regional, state and wider levels
- Economic development
- Maintenance of the cultural, economic, physical and social wellbeing of people and communities."

In May 2002, the Queensland Department of Public Works published a waste management strategic plan to establish "an integrated framework to minimise and manage waste in accordance with the principles of Ecologically Sustainable Development promoting efficient resource use."

The Queensland Department of Housing has initiated the "Toward Healthy and Sustainable Housing Research Project," which was an essential part of the Queensland Government's initiative and commitment to reduce greenhouse gas emissions.

In 2007, the ClimateSmart 2050 – Queensland Climate Change Strategy was launched to promote low-carbon future in the sectors of energy, industry, community, planning and building, primary industries and transport. The key initiative included developing new technologies, Smart Energy Savings Program, Renewable Energy Fund, Climate Smart Home Rebates, safe storage of carbon dioxide emissions, new planning standards for commercial buildings and so on. The Smart Housing Program was initiated to construct 30 sustainable demonstration homes across the state to promote smarter approaches to house design.

In Western Australia, the Department of the Premier and the Cabinet had established a Sustainability Policy Unit within the Policy Office to develop the State Sustainability Strategy. For this strategy, a number of background papers were prepared, including one on Sustainable Building and Construction. In 2003, the Western Australian Government issued a State Sustainability Strategy. There were a wide range of strategies involving governance, natural resources, settlements and community.

Tasmania had a Resource Management and Planning System in place for sustainable development. The objectives of this system were:

- To provide for fair, orderly and sustainable use and development of air, land and water

- To promote the sustainable development of natural and physical resources and the maintenance of ecological processes and genetic diversity
- To encourage public involvement in resource management and planning
- To facilitate economic development in accordance with the above objectives
- To promote the sharing of responsibility for resource management and planning between the different spheres of the Government, community and industry in the State.

The two main legislations that supported the above system were: the Land Use planning and approval Act (1993) and the Environmental Management and pollution Control Act (1994). The primary objectives of these acts were pollution prevention, reuse and recycling of materials, clean production technology, waste minimisation and reduction of the discharge of pollutants and hazardous substance.

One of the key aspects of this report was the definition of "sustainable development" – defined as *managing the use, development and protection of natural and physical resources in a way, or at a rate, which enables people and communities to provide for their social, economic and cultural well-being and for their health and safety while*:

- Sustaining the potential of natural and physical resources to meet the reasonably foreseeable needs of future generations
- Safeguarding the life-supporting capacity of air, water, soil and ecosystems
- Avoiding, remedying or mitigating any adverse effects of activities on the environment.

Northern Territory had the Office of Environment and Heritage responsible for several environment protection strategies within the territory as well as nationally. However, there was no specific focus for sustainability provision in building construction sector. Some of the strategies of the Office of Environment and Heritage were:

- In 1992, NT Waste Minimisation and Recycling Strategy was released.
- In 1995, NT Waste Management and Pollution Control Strategy was endorsed by the state government, and a progress report was presented in the Legislative Assembly in 1998.
- National Ozone Protection Strategy under the Ozone Protection Act 1990.
- National Cleaner Production Strategy and National hazardous waste initiatives and strategies under the Waste Management and Pollution Control Act 1998.

The Office of Environment and Heritage administers the Environmental Assessment Act, the National Environment Protection Council (NT) Act, the Ozone Protection Act, the Environmental Offences and Penalties Act and the Waste Management and Pollution Control Act. The Office also has a Greenhouse Unit to monitor a broad range of greenhouse issues and advise the NT Government on related policies.

Australian Capital Territory (ACT): In 2001, the ACT enforced an energy rating scheme entitled "ACT House Energy Rating Scheme (ACTHERS)" that the new dwellings require a minimum four star rating. The rating scheme was assessed with FirstRate software, which made the rating system more exigent. This scheme was later repealed by Adoption of Planning Guidelines 2003.

The same year, new tools and assessment processes were adopted by the ACT Government for developments in ACT, known as Designing for High Quality and Sustainability. The tools and processes included:

- Site Analysis Guidelines
- Quality Design Indicators
- Residential Sustainability Index
- Design Review Panel
- Design Response Report
- Rewards and Incentives

Another notable action taken by the ACT Government was the launch of "No Waste by 2010" Waste management strategy in 1996. Amendments were made in the Building Act 1972 and Building Code of Australia (ACT Appendix) that require a waste management plan to be part of any application for demolition of a building.

In 2003, People Place Prosperity: a Policy for Sustainability in the ACT was launched. The policy described the sustainability principles that the ACT Government will incorporate into its systems and operations, including (ACT Government, 2009):

- Embedding sustainability within its decision-making processes
- Promoting sustainability to the wider community
- Developing partnerships for sustainability with the ACT community and
- Developing indicators and reporting regularly on progress

In 2007, the ACT Climate Change Strategy was launched. This strategy provided an overview of climate change science, the predicted impacts on the state and the government's vision and direction for responding to climate change.

2.5 Summary

This chapter discussed some of the key international developments as well as national developments by the federal, state and territory and local governments. Internationally, organisations such as the United Nations (UN), the Organisation for Economic Co-operation and Development (OECD) and International Standard Organisation (ISO) have a significant contribution to the development. The history of policy development can be called to the very first event of 1972 Stockholm conference, followed by many successive conferences, with the attendance of a large number of countries. The policies developed by these international organisations were discussed in detail. Australia has shown its commitment to international developments as well as the policy developed nationally with the support of federal and state governments. The National Greenhouse Strategy, NABERS and Emission trading scheme (ETS) can be identified as some of the major contributions by the federal government to sustainable development.

References

Ashe, B., Newton, P. W., Enker, R., Bell, J., Apelt, R., Hough, R., … Davis, M. (2003). *Sustainability and the building code of Australia*.

Australia, C. O. (1997). *Victorian statewide assessment of ecologically sustainable forest management*. Retrieved from Canberra.

Australia, C. O. (1998). *The national greenhouse strategy: Strategic framework for advancing Australia's greenhouse response*. Australian Greenhouse Office, Commonwealth of Australia Canberra.

Beer, T. (2003). Environmental risk and sustainability. In *Risk science and sustainability* (pp. 39–61). Springer.

Bellamy, A. J. (2006). Whither the responsibility to protect? Humanitarian intervention and the 2005 World Summit. *Ethics & International Affairs, 20*(2), 143–169.

Bulkeley, H., Jordan, A., Perkins, R., & Selin, H. (2013). Governing sustainability: Rio+ 20 and the road beyond. *Environment and Planning C: Government and Policy, 31*(6), 958–970.

Carassus, J. (2004). *The construction sector system approach: An international framework* (CIB report). International Council for Research and Innovation in Building and Construction.

Clémençon, R. (2012). Welcome to the Anthropocene: Rio+ 20 and the meaning of sustainable development. *The Journal of Environment & Development, 21*(3), 311–338.

COAG. (1997). *Principles and guidelines for national standard setting and regulatory action by Ministerial Councils and StandardSetting Bodies*. Retrieved from Canberra, Australia.

de Oliveira, J. F. G., & Trindade, T. C. G. (2018). Sustainabilty indicators. In *Sustainability performance evaluation of renewable energy sources: The case of Brazil* (pp. 45–62). Springer.

Ga, U. (2015). *Transforming our world: The 2030 agenda for sustainable development*. Division for Sustainable Development Goals.

Government, A. (1999). *Annual reports on operation of the act*. Retrieved from https://www.environment.gov.au/epbc/about/reports-and-statistics

Government, A. (2009). *People, place, prosperity: The ACT's sustainability policy*. Retrieved from Canberra, Australia. http://www.cmd.act.gov.au/__data/assets/pdf_file/0003/119730/people_place_prosperity.pdf

Greenhouse, C. (2007). Energy reporting group. 2006. *A national system for streamlined greenhouse and energy reporting by business*. Retrieved from https://parlinfo.aph.gov.au/parlInfo/search/display/display.w3p;query=Id%3A%22legislation%2Fems%2Fr2857_ems_b59a44da-8f2b-4342-881b-00b85f9a4007%22.

Grubb, M., Koch, M., Thomson, K., Sullivan, F., & Munson, A. (2019). *The 'Earth Summit' agreements: A guide and assessment: An analysis of the Rio'92 UN conference on environment and development* (Vol. 9). Routledge.

Hens, L., & Nath, B. (2003). The Johannesburg Conference. *Environment, Development and Sustainability, 5*(1), 7–39.

Jefferson, M. (2006). Sustainable energy development: performance and prospects. *Renewable Energy, 31*(5), 571–582.

Jörgens, H. (2012). *National environmental policies: A comparative study of capacity-building*. Springer.

Keeble, B. R. (1988). The Brundtland report: 'Our common future'. *Medicine and War, 4*(1), 17–25.

Kibert, C. J. (2001). Policy instruments for sustainable built environment. *Journal of Land Use and Environmental Law, 17*, 379.

Lafferty, W. M., & Eckerberg, K. (2013). *From the Earth Summit to Local Agenda 21: working towards sustainable development*. Routledge.

Lehmann, S. (2006). Towards a sustainable city centre: Integrating ecologically sustainable development (ESD) principles into urban renewal. *Journal of Green Building, 1*(3), 83–104.

McInerney, L., Nadarajah, C., & Perkins, F. (2007). Australia's infrastructure policy and the COAG National Reform Agenda. *Economic Round-Up, 17*(Summer).

Newton, P. (2011). Consumption and environmental sustainability. *Urban Consumption*, 173–197.

Okpala, D. (1996). The second United Nations conference on human settlements (Habitat II). *Third World Planning Review, 18*(2), iii.

Protocol, K. (1997a). *Kyoto protocol*. UNFCCC Website. Available online: http://unfccc.int/kyoto_protocol/items/2830.php. Accessed on 1 Jan 2011.

Protocol, K. (1997b). United Nations framework convention on climate change. *Kyoto Protocol, Kyoto, 19*, 497.

Sitarz, D. (1993). *Agenda 21: The earth summit strategy to save our planet*. Earth Press.

Spangenberg, J. H. (2002). Institutional sustainability indicators: an analysis of the institutions in Agenda 21 and a draft set of indicators for monitoring their effectivity. *Sustainable Development, 10*(2), 103–115.

Sullivan, E. T. (1972). The Stockholm conference: A step toward global environmental cooperation and involvement. *Indiana Law Review, 6*, 267.

Summit, M. (2000). United Nations Millennium Declaration. In United Nations (Ed.). Retrieved from https://www.ohchr.org/EN/ProfessionalInterest/Pages/Millennium.aspx Accessed on 14-03-2019.

Territory, L. A. f. t. A. C. (1997). Legislative Assembly for the ACT: 1997. Retrieved from https://www.hansard.act.gov.au/Hansard/1997/week12/3928.htm

UNDESA. (2007). Retrieved from https://sustainabledevelopment.un.org/index.php?page=view&type=400&nr=107&menu=1515

UNEP. (2002). *Melbourne principles for sustainable cities*. Retrieved from Nairobe:

Von Frantzius, I. (2004). World Summit on sustainable development Johannesburg 2002: A critical analysis and assessment of the outcomes. *Environmental Politics, 13*(2), 467–473.

Wheeler, N. J. (2005). A victory for common humanity-The responsibility to protect after the 2005 World Summit. *Journal of International Law & International Relations, 2*, 95.

Chapter 3
Climate Change and Built Environment

3.1 What Is Climate Change?

3.1.1 Climate System

The climate system is consisting of the atmosphere, land surface, snow and ice, ocean and other bodies of water, and living things, which dynamically interact with each other. The system changes from time to time, affected by both internal dynamic characters and external forces involved with natural and human-made variations (Pachauri et al., 2014; Solomon et al., 2007; Stocker et al., 2013). Figure 3.1 shows the main components of our climate system and their interaction between each other. Some of the basic terms to represent our climate system were defined by IPCC (Solomon et al., 2013) as:.

- Climate: Climate concerns the status of the entire earth system, including the atmosphere, land, oceans, snow, ice, and living things that serve as the global background conditions that determine weather patterns.
- Cryosphere: The component of the *climate system* consisting of all snow, ice, and *frozen ground* (including *permafrost*) on and beneath the surface of the earth and ocean.
- Hydrosphere: The component of the *climate system* comprising liquid surface and subterranean water, such as oceans, seas, rivers, freshwater lakes, underground water, etc.
- Biosphere (terrestrial and marine): The part of the earth system comprising all *ecosystems* and living organisms in the *atmosphere*, on land (terrestrial biosphere), or in the oceans (marine biosphere), including derived dead organic matter such as litter, soil organic matter, and oceanic detritus.

© Springer Nature Switzerland AG 2021
X. Wang, S. Ramakrishnan, *Environmental Sustainability in Building Design and Construction*, https://doi.org/10.1007/978-3-030-76231-5_3

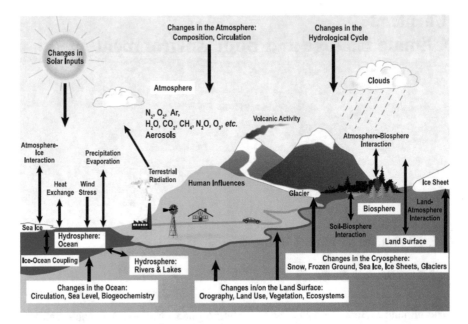

Fig. 3.1 Global climate system. (IPCC 2007, Solomon et al. (2007))

3.1.2   Climate States

Temperature, precipitation (e.g., rainfall and snow), and wind are normally applied to describe the state of the climate system, in terms of their mean and variability over a period, such as a month and year. The climate state is fundamentally determined by the earth's and global dynamic energy balance, which can be demonstrated by Fig. 3.2:

The mean energy rate of solar irradiation from the space, on the surface normal to the sun's rays beyond the earth's atmosphere, is about 1367 W/m^2, which is affected by the distance between the earth and the sun. The dynamic energy balance is theoretically maintained in terms of:

{Absorbed by Earth's Surface} =

+ {Solar Irradiation from the Space}
− {Re and Em from Clouds to the Space}
− {Re and Em from Atmo Gases to the Space}
− {Re and Em from Aerosol to the Space}
− {Re and Em from Earth Surface to the Space}
− {Absorbed by Atmosphere Gases}
− {Absorbed by Clouds}
− {Absorbed by Aerosol}.

where Re, Reflection; Em, Emission; Atmo, Atmosphere.

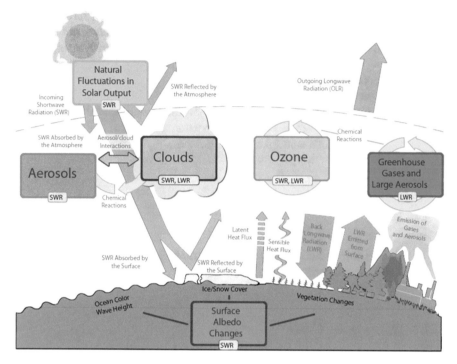

Fig. 3.2   Global dynamic energy balance (IPCC 2013, Stocker et al. (2013))

3.1.3   Greenhouse Effect

The history of the knowledge on greenhouse effect can be called back to the early nineteenth century. In 1824, the French mathematician *Joseph Fourier* gave the concept of "greenhouse effect" as the "surface heat on Earth was maintained by the atmosphere—otherwise the planet's orbit was too remote from the sun for a temperature that could support life."

Later in 1859, the Irish physicist *John Tyndall* observed that the atmospheric carbon dioxide (CO_2), methane (CH_4) and water vapor (H_2O) as key factors in maintaining temperature despite their tiny percentage of the total atmosphere.

In 1896, Swedish chemist Svante Arrhenius made a simple calculation to show that CO_2 accumulating in the atmosphere could increase the earth surface temperature roughly by 5 °C. It has now become a major field of research and a major driving force behind the sustainable development agenda. In 1925, the American statistician Alfred James Lotka described "anthropogenic climate change" as a major cause of current climate variations.

Greenhouse effect can be described as the process in which the absorption of infrared radiation by the atmosphere warms the earth. The change in the composition of atmospheric gases leads to energy rebalance. For example, the increase in GHG, which is mostly related to H_2O and CO_2, along with others including CH_4, nitrous oxide (N_2O), ozone (O_3), and several others, may reduce the infrared radiation through the atmosphere leading to the increase in the temperature of the earth's surface and the lower atmosphere. The term "greenhouse effect" may refer to natural greenhouse effect or enhanced (anthropogenic) greenhouse effect, which are due to naturally occurring greenhouse gases and the gases emitted as a result of human activities, respectively (Parry et al., 2007).

Greenhouse gases are those constituents in the atmosphere, that absorb and emit radiation at specific wavelengths within the spectrum of infrared radiation. The primary GHGs in the earth's atmosphere include:

– H_2O,
– CO_2,
– N_2O,
– CH_4 and
– O_3.

Note: The Kyoto Protocol also considers sulfur hexafluoride (SF_6), hydrofluorocarbons (HFCs), and perfluorocarbons (PFCs) as greenhouse gases.

3.1.4   Sources of GHG

- CO_2:

  – Fossil fuel use in transportation,
  – Building heating and cooling,
  – Manufacture of cement and other goods,
  – Natural processes such as the decay of plant matter, and
  – Deforestation releasing CO_2 and reducing its uptake by plants.

- CH_4 (*growth rates decreased over the last two decades*):

  – Human activities related to agriculture, natural gas distribution, and landfills and
  – Natural processes that occur, for example, in wetlands.

- N_2O:

  – Human activities such as fertilizer use and fossil fuel burning,
  – Natural processes in soils, and
  – Oceans.

- Water vapour – An important and most abundant greenhouse gas in the atmosphere.

  – The direct influence from human activities and natural events are slim. Indirectly, the changing climate due to human activities have the potential to affect water vapour concentration substantially. For example, a warmer atmosphere contains more water vapour.

- Aerosols – These are a collection of solid or liquid airborne particles, typically between the particle sizes of 0.01 and 10 μm, residing in the atmosphere for at least several hours.
  - Naturally occurring compounds.
  - Human activities including the fossil fuel and biomass burning process that generates aerosols of sulphur compounds, organic compounds and black carbon (soot); surface mining and industrial processes.
- Ozone – The release of gases such as carbon monoxide, hydrocarbons and nitrogen oxide, which chemically react to produce ozone in the troposphere. Meanwhile, in the stratosphere, human activities destroy ozone and have caused the ozone hole over Antarctica.
- Halocarbon gas concentrations (The recent international regulations designed to protect the ozone layer has resulted in a decreasing concentration).

  - Human activities: chlorofluorocarbons (e.g. FC-11 and CFC-12), which were used extensively as refrigeration, presence in the atmosphere was found to cause stratospheric ozone depletion.

3.2 Climate Change

Climate has always been changing, often in cycles and on different timescales. Several ice ages have occurred in the last two million years, and there seems to be a 20-year drought cycles. The expansion of glaciers every 100,000 years could result from the earth's orbital shift, while the drought cycle could be caused by the sun spot cycles. Climate does not always change gradually; meteor impacts and sudden shifts in ocean current could cause rapid climate changes. In general, climate change is referred as a change in the state of the climate that can be identified by changes in the mean and/or the variability (such as standard deviations, the occurrence of extremes, etc.), of the climate on all spatial and temporal scales beyond that of individual weather events, and that persists for an extended period, typical decades or longer. It must be underlined that the climate change does not represent the change in weather. Weather observations are short-term relevant, and the variation in weather conditions on daily, monthly, or yearly basis would not evident the change in climate.

Climate ↑ Weather

Climate change may attribute natural internal processes or external forcings. Here, external forcing refers to any forcing agent outside the climate system causing a change in the climate system. These include, volcanic eruptions, solar variations, and anthropogenic changes in the composition of the atmosphere and land use change.

However, the latest studies reveal that the current concern is with *human-caused climate change* or the climate change caused by anthropogenic activities. Strong evidence has suggested that it is extremely likely that the earth's climate system changed on both global and regional scales since the preindustrial era, with some of these changes attributable to human activities (Stocker et al., 2013). Past emissions of fossil fuels, cement production, and land use may have contributed the change, all of which are more or less related to construction.

Human-caused climate change has led to the variations in following constitutions in our atmosphere:

It has primarily resulted in changes in the amounts of *GHGs* in the atmosphere:

- CO_2,
- CH_4,
- N_2O, and
- The halocarbons (a group of gases containing fluorine, chlorine, and bromine);

also changes in

- Small particles (aerosols) and
- Land use;

directly or indirectly by socioeconomic processes such as

- Land use change (forestry to agriculture, agriculture to urban area),
- Land-cover modification (ecosystem degradation or modification).

3.3   Global Warming

The latest observation and analysis indicate the likely trend in global warming in terms of (Stocker et al., 2013):

- Global mean surface temperatures have risen. The rate of warming was observed to be almost double in the last 50 years, compared to last 100 years;
- Land regions have warmed at a faster rate than oceans;
- Changes in temperature extremes are also consistent with warming of the climate; and
- The frequency of heavy precipitation events have been substantially increased, consistent with a warming climate and observed significant increasing amounts of H_2O in the atmosphere.

Most scientists attribute the global warming in the last 50 years to the increased GHG level in the atmosphere. Six GHGs that contribute to the warming of the atmosphere are CO_2 (64%), CH_4 (19%), HFCs and PFCs (11%), N_2O (6%), and SF_6. The burning of fossil fuels is the largest contributor (5–5.5 billion tons of CO_2/year). About 47% of the CO_2 emissions stay in the atmosphere; 28% are absorbed into the ocean and 25% into vegetation.

Emissions are reported in tons of CO_2 equivalent (CO_2-eq) to account for the differences in global warming potential of different GHGs as shown in Table 3.1:

Table 3.1 Emission from various GHGs

GHG	CO_2 equivalent
CO_2	1
CH_4	28
N_2O	265
HFC	4,660-13,900
PFC	4-12,400
SF_6	23,500

3.4   Carbon Cycle

On our earth, there are many substances moving in a cyclic pathway through our biosphere, atmosphere, lithosphere, and hydrosphere. These are called biogeochemical cycles for the chemical elements of calcium, carbon, hydrogen, oxygen, nitrogen, phosphorus, sulfur, water, etc. Among these cycles, carbon cycle is important as all living organisms are comprised of carbon and its components. Carbon cycle can be defined as a biogeochemical cycle by which carbon is exchanged between the biosphere, atmosphere, lithosphere, and hydrosphere of the earth. The major molecules of carbon cycles are CO_2, carbon monoxide, CH_4, calcium carbonate, and many others. Figure 3.3 shows the exchange of carbon elements between different parts of the earth.

Among the different molecules, CO_2 is important as it regulates the earth's surface temperature, enabling the survival of all lives on the earth. The CO_2 circulates between the atmosphere, land biosphere and oceans. The atmospheric CO_2 removal involves a range of processes occurring at significantly large timescales. About 50% of a CO_2 increase will be removed from the atmosphere within 30 years, and a further 30% will require a few centuries for removal. The remaining 20% may stay in the atmosphere for many thousands of years.

Figure 3.4 shows the contribution of CO_2 to the atmosphere. Ocean uptaking of CO_2 was estimated to be 2.2 ± 0.5 GtC/yr between the 1990s and the first 5 years of the twenty-first century. Ocean CO_2 uptake has lowered the average ocean pH (increased acidity) by approximately 0.1 since 1750, lowering the CO_2 uptaking. Consequently, the marine ecosystems is affected by reduced calcification by shell-forming organisms and, in the longer term, the dissolution of carbonate sediments (Fig. 3.5).

Fires from natural causes and human activities may also change CO_2 in the atmosphere, by releasing considerable amounts of radioactively and photochemically active trace gases and aerosols. If fire frequency and extent increase with a changing climate, a net increase in CO_2 emissions is expected during this fire regime shift. A best estimation of the atmospheric temperature increase due to the increase in CO_2-eq is shown in Table 3.2.

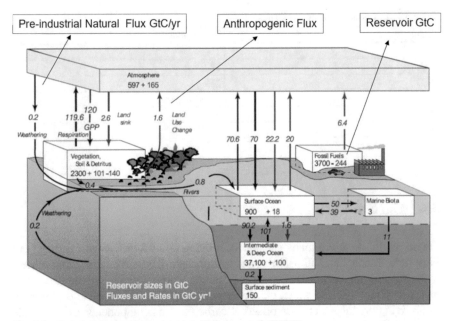

Fig. 3.3 Global carbon cycle on the earth (Solomon et al. (2007))

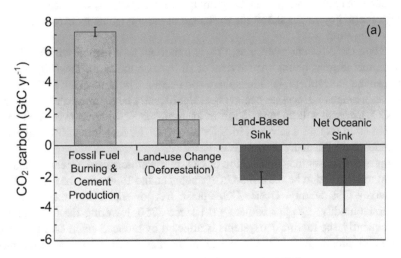

Fig. 3.4 Contribution of CO_2 to the atmosphere (Solomon et al. (2007))

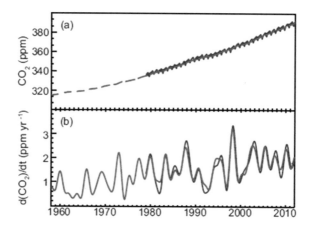

Fig. 3.5 CO_2 concentration in the atmospheric air (Stocker et al., 2013)

Table 3.2   Estimation of the range of temperature variation with CO_2-eq (Meehl et al., 2007)

Equilibrium CO_2-eq (ppm)	Temperature increase (°C)		
	Best estimate	Very likely above	Likely in the range
350	1.0	0.5	0.6–1.4
450	2.1	1.0	1.4–3.1
550	2.9	1.5	1.9–4.4
650	3.6	1.8	2.4–5.5
750	4.3	2.1	2.8–6.4
1000	5.5	2.8	3.7–6.3
1200	6.3	3.1	4.2–9.4

3.5   Climate Projection

The IPCC defines Climate change projections as the projection of the climate system's response to emission or concentration scenarios of GHGs and aerosols or radiative forcing scenarios, based upon simulations by climate models. Climate projections ($\neq$ prediction) are determined by the changes in heat energy in the earth system, caused by the increasing intensity of the greenhouse effect, determined by the amount of CO_2 and other GHGs in the atmosphere (Collins et al., 2013).

Climate change projections are produced by computer models of the climate system, known as Global Climate Models (GCMs). The GCMs are described as "numerical representation of the climate system based on the physical, chemical, and biological properties of its components, their interactions, and feedback processes and accounting for all or some of its known properties" (Pacheuri et al., 2014). The reliability of these models were justified by IPCC as follows:

- A credible quantitative estimates of future climate change can be obtained by climate models, particularly at continental scales and above.

- The confidence in projection comes from the foundation of the models in accepted physical principles and from their ability to reproduce observed features of current climate and past climate changes.
- Some climate variables (e.g., temperature) provide high confidence in estimation than others (e.g., precipitation).

There are still uncertainties and disagreement about the models and emission scenarios. The reasons include the following: (i) the potency of CO_2 in warming up the atmosphere, (ii) atmospheric CH_4 may be declining, (iii) the role of H_2O, and (vi) the role of negative feedback mechanism such as cloud cover and plant growth. The dominant view, however, is that global warming is a serious problem. The precautionary principle requires actions to be taken to avoid disaster despite the lack of scientific certainty.

3.6   Global Climate Change—Observations

The observations that evident global climate change can be divided into four major categories:

- Temperature,
- Precipitation,
- Sea level, and
- Extreme events.

When expressing the variations in climate parameters, a statistical term of anomaly is often used. Anomaly is defined as the variation in property from its long-period average value for the location concerned. For example, when the maximum monthly temperature for January in Sydney was 1.5 °C higher than the long-term average for this month, the anomaly will be +1.5 °C. The current international standard is to use the 30-year average from 1961 to 1990 as the long-term average.

Temperature Observations
Figure 3.6 shows observed annual anomalies of global land surface air temperature relative to 1961–1990 mean value. The global mean surface temperatures have risen by 0.74 °C ± 0.18 °C when estimated by a linear trend over the last 100 years. The increase rate over the last 50 years is 0.13 °C ± 0.03 °C per decade, almost the double of the rate of 0.07 °C ± 0.02 °C per decade over the last 100 years. Furthermore, from the observations of land and sea temperature trend, the following conclusions were dwarn by IPCC:

- Surface temperatures have risen globally, with a significant variation regionally.
- Concerning the global average, warming in the last century is classified into two phases, substantial variation (0.35 °C) from the 1910 to 1940s and more strongly (0.55 °C) from the 1970s to the present.

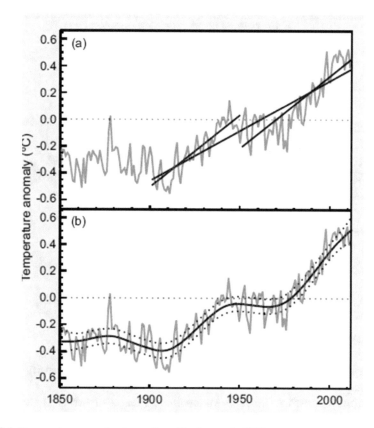

Fig. 3.6 Temperature anomaly observations (Stocker et al., 2013)

- A high rate of warming has taken place during the last 25 years, revealed by the frequent warmest years in the recent past; for instance, 11 of the 12 warmest years on record have occurred in the past 12 years. (2007).
- Apart from the land surface temperature rise, the confirmation of global warming also comes from warming of the oceans, glaciers melting, rising sea levels, diminished snow cover in the Northern Hemisphere and sea ice retreating in the Arctic.

Observation—Precipitation

The observed global precipitation anomaly trend over the 106-year period (1901–2005) is shown in Fig. 3.7. The figure 3.7 shows that the global precipitation anomaly is statistically insignificant. However, it must be stressed that global trend does not necessarily similar for regional trends. The regional trend on the observed precipitation during the period 1951–2010 compared with the 1901–2010 period is shown in Fig. 3.8. The observations of regional trend has led to draw following conclusions by IPCC:

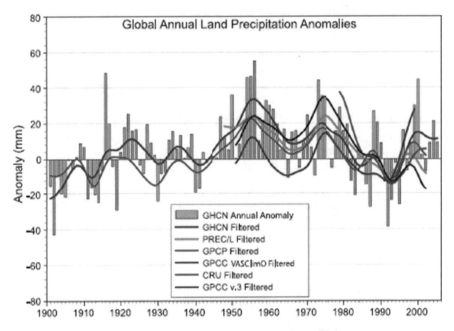

Fig. 3.7 Global precipitation anomaly observations (Solomon et al., 2007)

- The regional changes are occurring in the amount, frequency, intensity, and type of precipitation.
- While the southern Africa, the Mediterranean and southern Asia experiences drier environment, a significantly wetter seasons were observed in North and South America, northern Europe and northern and central Asia.
- The increases in heavy precipitation events have been observed in many places – Due to the increased water vapour in the atmosphere arising from the warming of the world's oceans, especially at lower latitudes.
- More precipitation now falls as rain rather than snow in northern regions.
- The events of both droughts and floods have increased in some regions.

Observation—Sea Level

The global mean sea level change from two major processes:

- Thermal expansion due to the sea temperature increase and
- The exchange of water between glaciers, oceans and other reservoirs, including glaciers and ice caps, ice sheets, other land water reservoirs.

Figure 3.9 shows the sea level rise trend from 1880 to 2005 as an anomaly of 1961 to 1991 period. Global mean sea level has been rising with the average rate of 1.7 ± 0.5 mm/year for the past century (or average sea level rise = 0.12–0.22 m), 1.8 ± 0.5 mm/year from 1961 to 2003, and 3.1 ± 0.7 mm/year from 1993 to 2003.

Further observations in sea level changes include the decrease in snow/ice cover in most regions, especially in spring and summer. The annual mean arctic sea ice

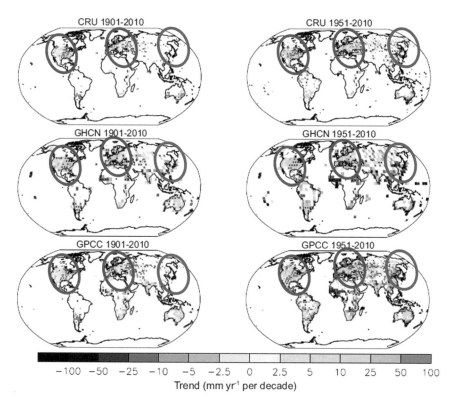

Fig. 3.8 Trends in annual precipitation over land for 1901–2010 (left-hand panels) and 1951–2010 (right-hand panels). White areas indicate incomplete or missing data. Black plus signs (+) indicate grid boxes where trends are significant (i.e., a trend of zero lies outside the 90% confidence interval) (Stocker et al., 2013)

extent declined 2.7 ± 0.6% per decade since 1978. The loss of glaciers and ice cap is about 0.50 ± 0.22 mm/year in sea level equivalent between 1961 and 2004. The oceans are also warming. Over the period of 1961–2003, global ocean temperature has risen by 0.10 °C from the surface to a depth of 700 m.

Observation—Extreme Events

The extreme event is defined by IPCC as "an event that is rare within its statistical reference distribution at a particular place". Definitions of "rare" vary, but an extreme weather event would normally be as rare as or rarer than the tenth or 90th percentile". Extreme weather events typically include heatwaves, cyclones, floods and droughts, etc.

Key issues of extreme events:

– Formation regions,
– Frequency,

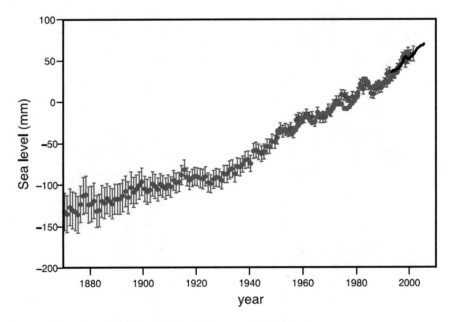

Fig. 3.9 Global sea level rise anomaly (Solomon et al., 2007)

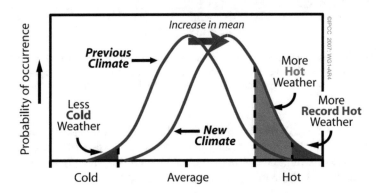

Fig. 3.10 Trend of extreme events as an increase in mean properties (Solomon et al., 2007)

– Intensity,
– Duration

In built environment, the damage from extreme events depends more on human factors than natural variability, including people, vulnerability, whether they place themselves in harm's way, and their resilience to extreme events through the variations in building codes and standards. The Fig. 3.10 shows the relationship between the mean and extreme of a climate event. The Fig. 3.10 explains an important phenomenon that the slight increase in the mean parameter shifts the entire profile toward right, significantly increasing the probability of the occurrence of extreme

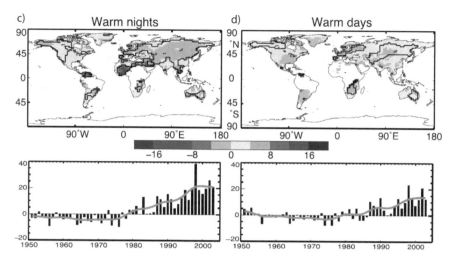

Fig. 3.11 Trends (day/decades) and warm nights/days of temperature at tenth percentile

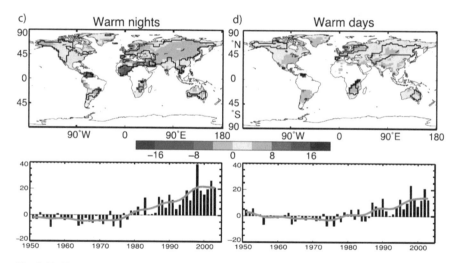

Fig. 3.12 Trends (day/decades) and cold nights/days of temperature at tenth percentile

event. Therefore, when considering the temperature as a climate parameter in this scenario, the slight increase in mean temperature will induce more hot weather with the increase in the highest recorded hot weather.

Figures 3.11 and 3.12 show the trends on the increase of warm nights/days and decrease of cold nights/days on a global level. Furthermore, the following observations were noted by IPCC on extreme events:

• Since 1950, the number of heat waves has increased with the consequences of widespread increases in the numbers of warm nights.

- The frequency of heavy daily precipitation events have increased leading to the flooding.
- The extent of regions affected by droughts has also increased, as precipitation over land has marginally decreased, while warmer conditions have caused increased evaporation.
- Tropical storm and hurricane events have substantially increased in intensity and duration since the 1970s, however, the frequency vary considerably from year to year.

3.7   Climate Projection

Emission Scenarios

Inter-governmental Panel on Climate Change (IPCC) (Collins et al., 2013) have established that climate change caused by the enhanced greenhouse effect is already occurring and future change is inevitable. In order to project the future climate variations and to compare them with the current trend, the scenario standardization was introduced. It is well known that the GHG emissions are the product of very complex dynamic systems, and the driving forces for GHG emissions are the technological changes, demographic development and socioeconomic development. In this regard, the historical development of climate projection scenarios can be called back to emission scenarios reported in the Special Report on Emisison Scenarios (SRES).

This report considers the effect of different driving forces to define the scenarios based on the strength of each driving forces. Scenarios are an appropriate tool to analyse how driving forces may influence future emission outcomes and to assess the associated uncertainties. To enable coordinated studies of climate change, climate impacts, and mitigation options and strategies with the same reference value, 40 emission scenarios were reported by IPCC.

Emissions scenarios characterises the release of greenhouse gases, aerosols, and other pollutants to the atmosphere as well as the information on land use and land cover in future, providing essential inputs to the climate models. They are based on the patterns of driving forces such as technological development, economic development and population growth. It must be underlined that the emission scenarios do not track the 'short term' fluctuations such as the variation in oil prices, business cycles etc. They are focused on long term trends (Council, 2011). The IPCC SRES emission scenarios are given in Table below. Among the different SRES, A1FI scenario is considered as an extremely high GHG emission scenario, A1B is a medium emission scenario, and A1T is a low emission scenario.

Emission scenarios and the driving forces (reproduced from (Stewart, Wang, & Nguyen, 2011)

A1	Very rapid economic growth, a global population that peaks in mid-century and declines thereafter, and the rapid introduction of new and more efficient technologies, substantial reduction in regional differences in per capita income	F1	Fossil-intensive
		T	Nonfossil energy
		B	A balance across all sources
A2	A very heterogeneous world, and preservation of local identities, continuous increase in population, regionally oriented economic development, more fragmented per capita economic growth and technological change		
B1	Same population trend as A1, rapid change in economic structures toward a service and information economy, reductions in material intensity, the introduction of clean and resource-efficient technologies		
B2	Emphasis on local solutions to economic, social; and environmental sustainability; continuous increase in global population at a rate lower than A2, intermediate levels of economic development, and less rapid and more diverse technological change than in B1 and A1		

With the development of SRES of climate, simulations were conducted to project the future climate conditions considering different scenarios. However, before projecting the future climate parameters, the developed numerical models should be validated with the observations. Therefore, the simulations on temperature anomaly for the past 105 years (1900–2005) were conducted and compared with the observations (Fig. 3.13). It can be seen that the simulated results well follow the observations on the temperature anomaly (although none of the models were perfectly fitting with the observations).

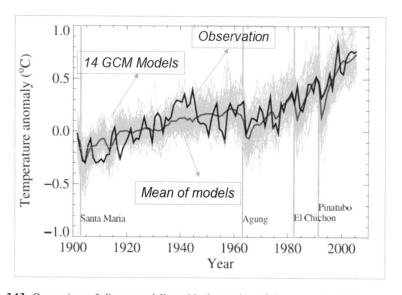

Fig. 3.13 Comparison of climate modeling with observations (Solomon et al., 2007)

In addition, the developed climate models were used to see the effect of anthropogenic forces in the temperature anomaly during past 105 years. The simulated temperatures with and without anthropogenic effects were compared with the observed temperature as shown in Fig. 3.14. It can be observed that the simulations that include only natural forcing are failed to follow the observed trend and they do not simulate the warming observed over the last three decades. On the other hand, simulations that incorporate anthropogenic forcing, including increasing GHG concentrations and the effects of aerosols, and that also incorporate natural external forcing, provide a consistent explanation of the observed temperature record.

With the comparison for past climate, the developed models were used to project the future climate conditions. Figure 3.15 shows the projected temperature anomaly for next 20 years for different scenarios. On the other hand, the sea level projections are also given in Fig. 3.16. The multimodal projections indicate that the temperature variations in 2030 could be from 0.5 to 2.0 °C increasing depending on the scenario, compared with the period of 1961–1990. On the other hand, the sea levels are projected to be risen by 0.05–0.35 m in 2030, compared with the present sea level. In addition to these climate variations, the simulations on the extreme events indicate the followings, as given in IPCC (Pachauri et al., 2014):

- It is very likely that heat waves will be more intense, more frequent, and lasting for longer durations in a future warmer climate. Subsequently, cold extremes are projected to decrease significantly.
- Precipitation generally increases in the areas of regional tropical areas and high latitute areas, with a decreasing trend in the subtropics.
- The intensity of precipitation events is projected to increase in tropical and high latitute areas, leading to flash flooding events.
- As the climate warms, the snow cover and sea ice extent will decrease, while the glaciers and ice caps lose mass, leading to the increase in summer melting over winter precipitation increases.
- Increasing atmospheric CO_2 concentrations leads directly to increasing acidification of the surface ocean.
- There is a likely increase of peak wind intensities and notably with increased near-storm precipitation in future tropical cyclones.

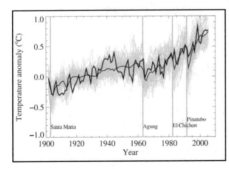 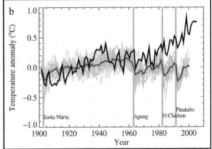

Fig. 3.14 Assessment of anthropogenic forces in the current climate (**a**) with anthropogenic forces (**b**) without anthropogenic forces (Solomon et al., 2007)

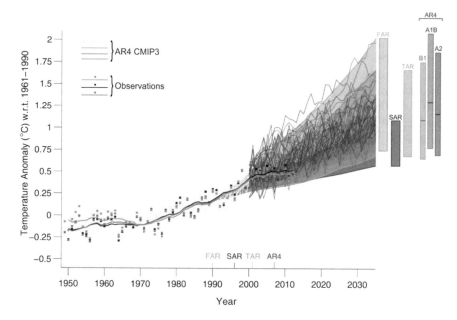

Fig. 3.15 Projected temperature anomaly for 2030

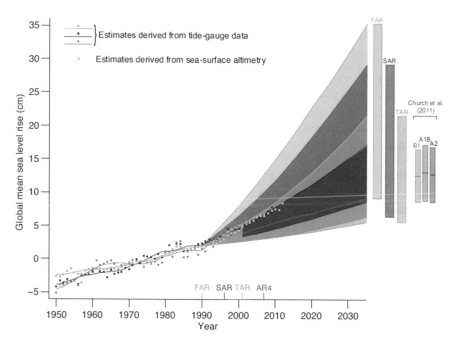

Fig. 3.16 Projected sea level rise for 2030

Over time, the climate modelling has also been significantly evolved. The latest IPCC report (IPCC AR6 report), with the collaboration of multiple communities, have introduced a framework known as Shared Socioeconomic Pathway (SSP)–Representative Concentration Pathway (RCP) framework. This framework considers a set of alternative socioeconomic development pathways (SSPs) and atmospheric concentration pathways (RCPs) to understand their associated climate change outcomes. In this framework, SSPs focus on societal factors such as demographics, human development, economic growth, inequality, governance, technological change and policy orientations. On the other hand, RCPs are pathways of the GHG concentrations over time as per emission scenarios.

3.8   Impact of Climate Change on Built Environment

The likely consequence for our human settlement due to climate change is the increased frequency of occurrences of natural disasters such as storms, floods, cyclones, and bushfires. These extreme events are causing a lot of damages to buildings. Although a level of uncertainty currently surrounds the magnitudes and implications of the changes, but the impacts on building construction are likely in the following areas:

- Bushfires—bushfire events are accelerated by the decrease in relative humidity, and increase of temperature and hot days (Lucas and Hennessy).
- Flooding— the flooding events might become more frequent. A study found that climate change as a result of GHG emission has increased the risk of flood events by more than 20% and in two out of three cases by more than 90% (Pall et al., 2011).
- Drought— the possible consequences are erosion, vegetation loss, and subsidence that might affect the foundations of buildings.
- High winds—Strong tropical cyclones and increased storm intensities may result in a requirement of higher design wind speeds (Wang & Wang, 2009a, b).
- To adapt to the change, strength and durability of structures may have to be increased. In addition, more airtightness may be required.
- Hot days—Increased hot days may lead to the change of heating ventilation and air conditioning energy consumption pattern (X. Wang et al., 2010) and also cause indoor environment (Huang et al., 2011; Kershaw et al., 2011)and health issues (Huang et al., 2011). Ecological systems may also be affected.

3.9   Coping with Climate Change: Mitigation and Adaptation

To cope with climate change, there are two approaches—*mitigation* and *adaptation*, essentially intended to reduce the adverse effects of human activities to our environment and the impact of worsened environment to human settlement. While mitigation aims to reduce anthropogenic impact on climate in medium-to-long term with global consequences, adaptation is the process or adjustment to reduce the adverse effects of climate change, generally in short-to-medium term with relatively local consequences. Properly designed climate change responses can be an integral part of sustainable development, and the two can be mutually reinforcing. It is generally considered that effective response to climate change must combine both mitigation to avoid the unmanageable and adaptation to manage the unavoidable.

3.9.1   Mitigation to Climate Change

The IPCC states Climate Change Mitigation (III, 2014) is the application of technological change and substitution to reduce resource inputs and emissions per unit of output. More specifically, it is an anthropogenic intervention to reduce the anthropogenic forcing of the climate system, by reducing the emission and enhancing the carbon sinks (a process to remove CO_2 from atmosphere). The anthropogenic interventions require social, economic, and technological policies that would produce an emission reduction. Meanwhile, we should also have implementing policies to reduce GHG emissions and enhance CO_2 sinks.

   In a review conducted by Stern Review (Stern et al., 2006), it was found that the costs of mitigation to avoid dangerous climate change could be limited to 1% of global gross domestic product (GDP) each year, however, the inaction might lead to losing 5–20% of global GDP per year and possibly more, as shown in Fig. 3.17.

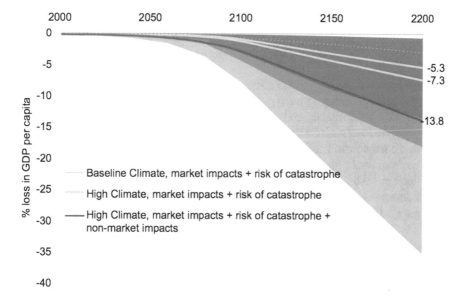

Fig. 3.17 Economic cost by mitigation and inaction (Stern et al., 2006)

The United Nations Framework Convention on Climate Change (UNFCCC) has formed from the increasing concern about the climate change and the recognition that this as a global environmental problem that no single country can solve for itself (Yu, 2004). At the Rio Earth Summit, parties to the UNFCCC agreed to stabilize emissions of GHGs at the 1990 levels by the year 2000. Australia signed the UN Convention in June 1992. In December 2007, Australian government has decided to ratify the Kyoto Protocol that came into effect on March 11, 2008. Australia's emissions are projected to be 108% of the 1990 level during the 2008–2012 period, as required under the Kyoto Protocol, which has been ratified by 178 countries. Australia has also set a target to reduce GHG emissions by 60% on the 2000 levels by 2050.

Department of Climate Change (DCC) provides the following figures calculated in accordance within UNFCCC inventory account provisions. Methodology used for the estimation of Australia's national GHG inventory has been detailed in a series reports. As reported by DCC, in Australia, the total net GHG emission in 2006 is 576 million tons CO_2-eq emission (or Mt. CO_2-e), that is, 4.2% above the 1990 level. More details in each sector are given below:

	Mt CO_2-e		%
	1990	2006	Variation
Energy	286.4	400.9	40%
Stationary use	*195.1*	*287.4*	*47.3%*
Transport	*62.1*	*79.1*	*27.4%*
Fugitive emission	*29.2*	*34.5*	*18.1*
Industrial processes	24.1	28.4	17.7%
Agriculture	86.8	90.1	3.8%
Land use	18.8	16.6	−11.4%
Waste	136.5	62.9	−53.9%
Forestry	0	−23.0	NA
Net emissions	552.6	576.0	4.2%

Stationary use includes combustion of solid, gaseous, and liquid fuels in energy industry, manufacturing industries, and construction and other energy use by the commercial, institutional, and residential equipment in each sector. Transport includes emission from road, rail, and domestic air and water transport. The fugitive emission comes from leaks during storage and transmission and also includes CH_4 emitted from coal mine seams. As seen, energy sector contributed the most GHG emissions at 69.6%. There were contributions from others, including agriculture 15.6%, land use and forestry (6.9%), industrial process 4.9%, and waste 2.9%.

Figure 3.18 indicates the variations of GHG emission in seven sectors from 1990 to 2006. Energy consumption was the main factor to drive up the emission. Therefore, reduction of energy demand would be critical to reduce GHG emission.

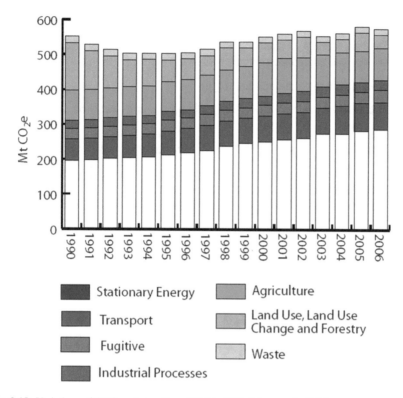

Fig. 3.18 Variations of GHG emission from 1990 to 2006 (Metz et al., 2007).

Emission Trading Scheme

One of the Australian Government's climate change strategies is the mitigation by reducing carbon pollution and adopting emission trading. the details of Emission Trading Schemes (ETSs) are as follows:

- The government has set a cap on emissions to be consistent with its long-term goal and international obligations. The cap will be turned into a number of "permits."
- Significant emitters need a "permit" for every ton of emission—the quantity of emission from each emitter will be monitored and audited.
- At the end of each year, each emitter need to surrender the "permit" for every ton of emission it produced in that year.
- "Permits" is available to trade as a commodity on the market and thus give incentive for industry to reduce emission.

Thus, ETSs put a price on emission throughout the economy and increased the cost of goods and services that are emission-intensive.

Mitigation Strategies in Energy Sector in Australia

Some of the mitigation strategies in the energy sector of Australia can be listed as follows:

- Energy efficiency, demand reduction, and distributed power generation system

  - Improved energy efficiency in buildings, smart energy management, and small-scale energy generation close to the point of use

- Renewable energy and nuclear energy

  - Solar-based power, wind power, biomass, hydropower, ocean energy, and nuclear energy

- Carbon capture and storage (CCS) and fossil fuel energy management

  - Fossil fuel–released CO_2 is captured and stored in stable geological formations such as deep underground, deep ocean, etc.
  - Common CCS technologies are gasification, postcombustion capture, mineralization, etc.

3.9.2   Adaptation to Climate Change

Climate Change Adaptation (Field, 2014) is defined as "initiatives and measures to reduce the vulnerability of natural and human systems against actual or expected climate change effects. Various types of adaptation exist, for example, anticipatory and reactive, private and public, and autonomous and planned". It can be implemented at different jurisdictional levels, such as federal, state, and local councils, and it can also be implemented at household and individual levels.

Adaptation is fundamentally achieved by adjusting natural and human systems to diminish system *vulnerability* to actual or expected climate stimuli by decreasing climate variability sensitivity and increasing adaptive capacity. In another form, adaptation can be considered as a measure or process to prevent the systems from exceeding adaptive capacity and to reduce the variation of the systems performance in response to per variation of a climate variable, such as temperature and rainfall.

The adaptive capacity can be related to wealth, technology, information skill, infrastructure, access to resources, management capability, organization, and governance. For example, the increase of access to finance, insurance, and infrastructure may enhance the adaptive capacity of community in response to climate change. Enhancement of adaptive capacity is critical in the development of adaptation strategies.

In general, an adaptation strategy aims to enhance the resistance and robustness of human and natural systems to possible impacts as a result of climate changes in a cost-effective and feasible way, also with consideration of social dimensions of distributing loss.

In relation to the timing of adaptation implementation, it could generally be categorized into:

- Anticipatory or preventive adaptation: Act before any impact occurs. For example, regulation in building codes and planning;

- Progressive adaptation: Act during any impact occurs. For example, emergency and disaster response; and
- Reactive or corrective adaptation: Act after any impact occurs. For example, reconstruction after disaster.

Similar types of adaptation were also identified by IPCC, which include:

- *Anticipatory* adaptation (or proactive): It takes place before the observation of climate change impacts;
- *Autonomous* adaptation (or spontaneous): It is triggered by ecological changes in natural systems; and
- Planned adaptation: It is the result of awareness of the need of policy decisions or measures.

Adaptation strategies have been investigated by all Australian governments since the National Climate Change Adaptation Programme announced in 2004. Department of the Environment and Heritage published a report of "Climate Change: Risk and Vulnerability," done by the Allen Consulting Group in 2005. The report takes a risk management approach to identifying the sectors and regions that might have the highest priority for adaptation planning.

The difference of adaptation strategy from mitigation is presented through the facts (Jones & Preston, 2006):

- They essentially deals with different parts of risk: mitigation strategy reduces the likelihood and magnitude of hazards by reducing the source or enhancing the sinks of greenhouse gas emissions; adaptation strategy reduces / manages the consequences of the impacts caused by climate–related hazards.
- Operating at different parts of the potential climate change envelope: mitigation works on the upper limit of the plausible range by reducing the likelihood of climate change risks; adaptation works with the lower limit of the plausible range by managing experienced or more probable climate change risks.
- Effective over different timescales: mitigation strategies enable long term benefits because of the delayed response of climate change; adaptations have benefits in the short to medium term, particularly they manage current climate risks.
- Effective at different spatial scales: the mitigation strategies reduces climate change associated risks at the global scale since the associated elements are well mixed in the atmosphere; adaptation is specific to local conditions.

3.10 Summary

This chapter introduced climate change as a major threat to the built environment and presented information on how the built environment is affected by changing climate. The definitions of climate systems, climate states, GHG emissions, and consequences to built environment are discussed. The global climate change has been studied with the observations that evidence the changing trend on temperature, precipitation, and sea levels in the past 200 years compared with the long-term average of 30 years. Extreme events have been identified as a major threat to the human and built environment with the increase in frequency and intensity of extreme with the changing climate. The climate projection to assist with projecting future climate was discussed along with the emission scenarios to support the projection. The climate projection using emission scenarios not only supports the changing climate in the future, but it is evident that the current climate is mainly caused by the anthropogenic actions. This means, if there are no human-induced climate change forces, the current climate conditions were much lower in terms of climate parameters such as GHG concentrations, temperature, etc. Finally, the methods to cope with the changing climate were discussed with the mitigation technologies and adaptation methods to the climate change.

References

Collins, M., Knutti, R., Arblaster, J., Dufresne, J.-L., Fichefet, T., Friedlingstein, P., ... Krinner, G. (2013). Long-term climate change: Projections, commitments and irreversibility. In *Climate change 2013-The physical science basis: Contribution of working Group I to the fifth assessment report of the intergovernmental panel on climate change* (pp. 1029–1136). Cambridge University Press.

Council, N. R. (2011). *Modeling the economics of greenhouse gas mitigation: summary of a workshop*: National Academies Press.

Field, C. B. (2014). *Climate change 2014–Impacts, adaptation and vulnerability: Regional aspects*. Cambridge University Press.

Huang, C., Vaneckova, P., Wang, X., FitzGerald, G., Guo, Y., & Tong, S. (2011). Constraints and barriers to public health adaptation to climate change: A review of the literature. *American Journal of Preventive Medicine, 40*(2), 183–190.

III, I. W. G. (2014). *Climate change 2014-mitigation of climate change: Summary for policymakers*. Intergovernmental Panel on Climate Change.

Jones, R., & Preston, B. (2006). *Climate change impacts, risk and the benefits of mitigation*. Paper presented at the A report for the Energy Futures Forum.

Kershaw, T., Eames, M., & Coley, D. (2011). Assessing the risk of climate change for buildings: A comparison between multi-year and probabilistic reference year simulations. *Building and Environment, 46*(6), 1303–1308.

Meehl, G. A., Stocker, T. F., Collins, W. D., Friedlingstein, P., Gaye, A. T., Gregory, J. M., . . . Noda, A. (2007). Global climate projections. Chapter 10.

Metz, B., Davidson, O., Bosch, P., Dave, R., & Meyer, L. (2007). *Climate change 2007: Mitigation of climate change*. Cambridge University Press.

Pachauri, R. K., Allen, M. R., Barros, V. R., Broome, J., Cramer, W., Christ, R., ... Dasgupta, P. (2014). *Climate change 2014: Synthesis report. Contribution of working groups I, II and III to the fifth assessment report of the Intergovernmental panel on climate change*. IPCC.

Pall, P., Aina, T., Stone, D. A., Stott, P. A., Nozawa, T., Hilberts, A. G., ... Allen, M. R. (2011). Anthropogenic greenhouse gas contribution to flood risk in England and Wales in autumn 2000. *Nature, 470*(7334), 382–385.

Parry, M., Parry, M. L., Canziani, O., Palutikof, J., Van der Linden, P., & Hanson, C. (2007). *Climate change 2007-impacts, adaptation and vulnerability: Working group II contribution to the fourth assessment report of the IPCC* (Vol. 4): Cambridge University Press.

Solomon, S., Manning, M., Marquis, M., & Qin, D. (2007). *Climate change 2007-the physical science basis: Working group I contribution to the fourth assessment report of the IPCC* (Vol. 4). Cambridge University Press.

Stern, N. H., Peters, S., Bakhshi, V., Bowen, A., Cameron, C., Catovsky, S., … Edmonson, N. (2006). *Stern Review: The economics of climate change* (Vol. 30). Cambridge: Cambridge University Press.

Stewart, M. G., Wang, X., & Nguyen, M. N. (2011). Climate change impact and risks of concrete infrastructure deterioration. *Engineering Structures, 33*(4), 1326–1337. doi:https://doi.org/10.1016/j.engstruct.2011.01.010

Stocker, T. F., Qin, D., Plattner, G.-K., Tignor, M., Allen, S. K., Boschung, J., … Midgley, P. M. (2013). Climate Change 2013. The Physical Science Basis. Working Group I Contribution to the Fifth Assessment Report of the Intergovernmental Panel on Climate Change-Abstract for decision-makers; Changements climatiques 2013. Les elements scientifiques. Contribution du groupe de travail I au cinquieme rapport d'evaluation du groupe d'experts intergouvernemental sur l'evolution du CLIMAT-Resume a l'intention des decideurs.

Wang, C., & Wang, X. (2009a). *Hazard of extreme wind gusts in Australia and its sensitivity to climate change.* Paper presented at the 18th world IMACS/MODSIM Congress.

Wang, C., & Wang, X. (2009b). *Hazard of extreme wind gusts to buildings in Australia and its sensitivity to climate change.* Paper presented at the 18th World IMACS/MODSIM Congress.

Wang, X., Chen, D., & Ren, Z. (2010). Assessment of climate change impact on residential building heating and cooling energy requirement in Australia. *Building and Environment, 45*(7), 1663–1682.

Yu, H. (2004). Global environment regime and climate policy coordination in China. *Journal of Chinese Political Science, 9*(2), 63–77.

Chapter 4
Energy and Carbon Emission

4.1 Introduction

This chapter focuses on examining the current trends of energy use and its impacts on the environment from a global perspective and in the Australian context. It first gives an overview of different energy sources, production and consumption and then discusses the relationship between energy per capita and human development as a mean of assessing the sustainable development. More importantly, the energy and carbon dioxide (CO_2) emission of building sector during the life cycle of buildings will be demonstrated in detail with the aid of carbon-accounting frameworks applied in construction. Various energy principles and technologies to improve building energy efficiency will also be proposed and discussed.

4.2 Energy Sources, Production and Consumption

4.2.1 Primary and Secondary Energy Sources

The numerous energy sources used in the world can be categorised as primary and secondary energy sources. Primary energy sources are those that can be directly used, as they are naturally available. Some of the examples of primary energy sources are:

- Solar energy, which is responsible for solar thermal energy, hydropower, wind, photovoltaic energy as well as the production of biomass energy
- Coal, oil, natural gas and wood sources
- Geothermal energy, originated from the molten core of the earth
- Nuclear energy, originated from the nuclei of atoms
- Tidal energy, originating from the gravitational attraction from the moon

© Springer Nature Switzerland AG 2021
X. Wang, S. Ramakrishnan, *Environmental Sustainability in Building Design and Construction*, https://doi.org/10.1007/978-3-030-76231-5_4

Among the different primary energy sources, the sun's energy is the dominant source of energy on the earth. Figure 4.1 compares the global primary energy sources from 1973 to 2018 as a pie chart representing the fraction of each primary energy source. It can be seen that total primary energy sources increased from 6097 Mtoe to 13,972 Mtoe during the past four decades (IEA, 2019). Also the renewable energy sources including biofuels and waste, hydro, nuclear and other energies have increased from 13.2% to 18.7%.

In Australia, the primary energy sources are rich in content compared with national energy needs. As reported, Australia has:

- Proved coal reserve of 147,435 million tonnes or 14% of total world coal reserve in 2018, ranked third, behind the United States (23.7%) and Russia (15.2%).
- Proved natural gas reserve of 2.4 trillion cubic metre or 1.2% of total world reserve.
- Proved oil reserve of 400 million tonnes or 0.2% of the proved world reserve.

Primary energy sources can be divided into renewable energy sources and non-renewable energy sources. Renewable energy sources are produced from geophysical or biological sources that are naturally replenished at the rate of extraction. For example, photovoltaic solar energy, wind energy, tidal energy, hydropower, etc. are renewable energy sources. The development of renewable energy has been advocated to reduce greenhouse gas (GHG) emission and mitigate climate change impacts, while it would also be a solution coping with the problem of peaking in oil supply. On the other hand, nonrenewable energy sources, such as fossil fuels (coal, oil and natural gas), are characterised by long regeneration times with the extraction rates much higher than the replenishment rates. For example, fossil fuels took

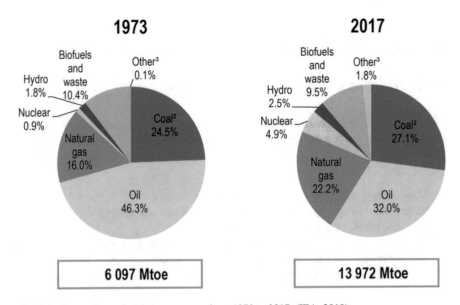

Fig. 4.1 Comparison of global energy use from 1973 to 2017. (IEA, 2019)

million years to form, and although still plenty of those sources are available, they are limited by increasing extraction rate and have a finite lifetime. The nuclear energy is generated by a controlled fission process of uranium-235 in nuclear reactor to heat the water (270 °C) into steam that drives electricity generator. It has been reported that the nuclear energy is not strictly a renewable energy because the uranium reserves are also finite.

The secondary energy sources are derived from the transformation of primary energy sources. For example, electrical energy that is derived from the conversion of mechanical energy, chemical energy or nuclear energy and petrol that is derived from the processing of crude oil are known as secondary energy sources. It must be underlined that the transformation of primary to secondary energy sources leads to an energy loss of up to 30%.

4.2.2 The Energy Consumption

The energy production and consumption as a means of available world reserve are given in Table 4.1. The unit is million tonnes for oil and coal and billion cubic metre for natural gasses. The trend of world energy consumption is also plotted in Fig. 4.2. Global energy consumption increased by 2.9% in 2018. Growth was the strongest since 2010 and almost doubled the 10-year average. The demand for all fuels increased, but growth was particularly strong in the case of gas (168 Mtoe, accounting for 43% of the global increase) and renewables (71 Mtoe, 18% of the global increase).

The production and consumption of fossil fuels, hydroelectric energy, and renewable energy are presented in Table 4.2. The share indicates the percentage of total production or consumption in the world. The higher consumption than production of oil implies that about 72% of consumption in 2018 was derived from import. In Australia, there are large reserves of nonrenewable energy, particularly coal and natural gas. These reserves are large compared with the national energy needs as shown in Figs. 4.3 and 4.4. The use of these resources may be constrained by environmental consideration rather than availability.

Table 4.1 Primary energy production, consumption and reserve (Dudley, 2018)

	Oil	Natural gas	Coal
Reserve	244,100 Mt	196,900 Bm³	1,054,782 Mt
Production	4474.3 Mt	3867.9 Bm³	3916.8 Mt
Consumption	4662.1 Mt	3848.9 Bm³	3772.1 Mt

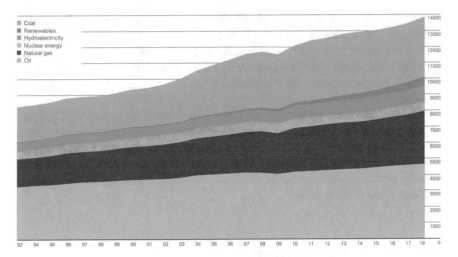

Fig. 4.2 Primary energy consumption by sources (Dudley, 2018; IEA, 2019)

Table 4.2 Energy production and consumption in Australia (Government, 2019)

	Production		Consumption	
	Quantity	Share	Quantity	Share
Oil	15.2 Mt	0.3%	53.4 Mt	1.1%
Natural gas	130.1 Bm³	3.4%	41.4 Bm³	1.1%
Coal	301.1 Mt	7.7%	44.3 Mt	1.2%
Hydroelectric energy	–	–	3.9 Mt	0.4%
Other renewables	–	–	7.2 Mt	1.2%

4.3   Energy per Capita and Human Development

In 2019, top ten nations of energy consumers are China, the United States, India, Russia, Japan, Canada, Germany, Brazil, South Korea and Iran. Australia is at the 17th position. The top ten nations' energy use per capita is given in Table 4.3, and Australia is ranked at the 13th position. To some extent, energy use per capita is an appropriate indicator related to sustainable development. Table 4.3 shows the energy consumption per capita in 2019.

It is interesting that there is no correlation between energy consumption and standard of living. Japan and Western Europe have high standard of living comparable with (or may be even better than) that of the United States, Canada and Australia and yet have far lower energy consumption rate. However, it must be noted that these countries export some of the energy in terms of either products or manufactured goods to other countries.

United Nations Development Programme (UNDP) Human Development Report (UNDP, 2007) found that human development index (HDI), which reflected several key social aspects related to sustainable development, including healthy life,

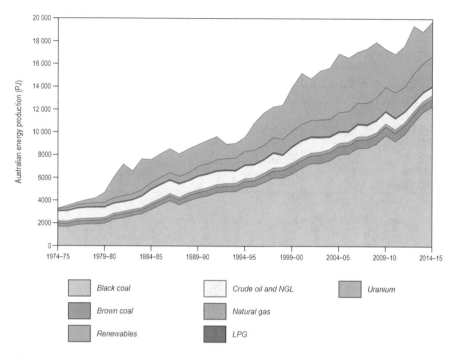

Fig. 4.3 Energy production in Australia from 1974–1975 to 2014–2015 (Government, 2019)

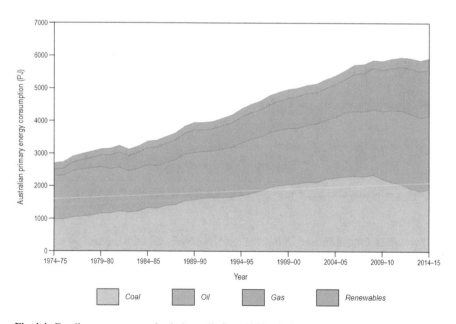

Fig. 4.4 Fossil energy consumption in Australia from 1975–1976 to 2014–2015 (Government, 2019)

Table 4.3 Energy use per capita in 2019 (Dudley, 2018)

Country	Energy per capita (gigajoule per capita)	Country	Energy per capita (gigajoule per capita)	Country	Energy per capita (gigajoule per capita)
Qatar	749.7	Kuwait	388.6	Oman	266.0
Iceland	696.4	Norway	370.6	South Korea	246.3
Singapore	633.0	Saudi Arabia	323.4	Australia	243.9
United Arab Emirates	492.3	The United States	294.8	Belgium	226.4
Trinidad and Tobago	465.5	Luxembourg	282.8	Turkmenistan	225.4

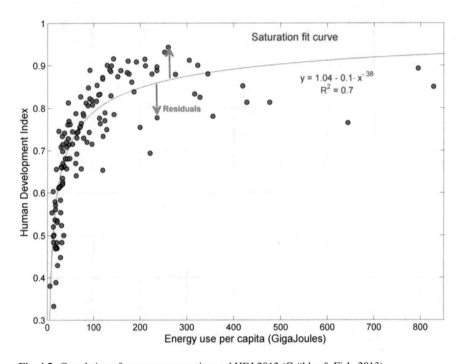

Fig. 4.5 Correlation of energy consumption and HDI 2012 (Grübler & Fisk, 2013)

knowledge or education and living standard, is correlated with the energy consumption per capita (tonnes of oil equivalent per capita). The latest developments on HDI with the per capita energy consumption are shown in Fig. 4.5.

4.4 Energy and CO_2 Emission of Construction Sector

As reported by Intergovernmental Panel on Climate Change (Solomon et al., 2007), direct emission from the building sector (excluding emissions from electricity use) was about 5 Gt CO_2 equivalent (CO_2-eq) in 2004. It reached 8.6 Gt when indirect emission from electricity use was considered. It is almost a quarter of the total global emission.

The Australian Greenhouse Office has sponsored various studies to provide projected levels of energy consumption and greenhouse gas emissions for the Australian residential and nonresidential building sectors. The results are summarised in Table 4.4.

A study on the energy consumption and CO_2 emissions of variety of buildings from cradle to grave with 50 years of lifespan has found that the energy consumption from these buildings is about 80–90% of the total primary energy and CO_2 emission, whereas the remaining 10% is a result of embodied energy and demolition (Chau et al., 2015).

Reduction of CO_2 emission can be implemented by the following measures:

- Reduce embodied energy in buildings, which is directed to the energy consumption in the manufacturing and transport of building materials and products.
- Reduce energy consumption, for example, via energy efficiency. It is directly related to energy demand for building operation, known as operating energy.
- Switch to low-carbon building construction and operation process, for example, through the use of renewable energy and low-carbon fuels.

Considering that the majority of CO_2 emission is related to building use rather than manufacturing, maintenance and demolition, building energy efficiency would be the most cost-effective way to reduce GHG emission of buildings. It should be mentioned that sustainable development may provide many options for GHG emission reduction, while sustainability embraces more broad domains, not only in environment, but also on social and economic aspect.

It is understood that reduction of GHG emission could be achieved in building sector, which is not only a benefit to the aversion of climate change, but also the achievement of sustainable development and economic goals. However, it may require strong push at policy level. Both technologies and policies that address

Table 4.4 Residential and nonresidential building energy use and CO_2 emission

	1990	2010	Increase
Energy use (PJ/year)			
Residential	270	379	40%
Nonresidential	151	289	91%
Greenhouse gas emission (Mt/year)			
Residential	48.6	56.7	17%
Nonresidential	32.2	62.8	94%

integrated solutions for mitigation and adaptation are important for the sustainable development under climate change.

4.5   Embodied Energy

Embodied energy is the energy consumed by all of the processes associated with the production and supply of building materials, components and structures and the construction of buildings. The energy embodied in the existing building stock in Australia is equivalent to about 10 years of the total energy consumption for the entire nation. Table 4.5 gives the embodied energy of some typical building materials. CO_2 emissions are highly correlated with the energy consumed in manufacturing building materials.

As seen from Table 4.5, the embodied energy per unit mass of materials used in building construction can be enormously different, from about 0.5 MJ/kg for air-dried sawn hardwood to 170 MJ/kg for aluminium. Apart from the embodied energy, other factors also affect environmental impact of materials, such as the variation in quantity of materials to perform same task, differing lifetimes of materials leading

Table 4.5 Embodied energy of typical building materials (Milne & Reardon, 2013)

Material	Embodied energy (MJ/kg)	Material	Embodied energy (MJ/kg)
Kiln-dried sawn softwood	3.4	Kiln-dried sawn hardwood	2
Air-dried sawn hardwood	0.5	Hardboard	24
Particle board	8	Medium-density fibreboard (MDF)	11
Plywood	10	Glue-laminated timber	11
Laminated veneer lumber	11	Plastics – general	90
Polyvinyl chloride (PVC)	80	Synthetic rubber	110
Acrylic paint	62	Stabilised earth	0.7
Imported dimensioned granite	14	Local dimensioned granite	5.9
Gypsum plaster	2.9	Plasterboard	4.4
Fibre cement	4.8	Cement	5.6
In situ concrete	1.9	Precast steam-cured concrete	2
Precast tilt-up concrete	1.9	Clay bricks	2.5
Concrete blocks	1.5	Autoclaved aerated concrete (AAC)	3.6
Glass	13	Aluminium	170
Copper	100	Galvanised steel	38

to intermediate replacements and different design requirements. For example, when comparing the brick veneer and fibre cement sheets as exterior walls for dwellings, bricks have high embodied energy and no structural role while they do not require a replacement; fibre cement sheets have lower embodied energy with similar structural and thermal performance, however, intermediate replacement is required during the lifespan of the building.

Materials such as concrete and timber have the lowest embodied energy intensities, but are consumed in very large quantities. On the other hand, steel has high embodied energy, however, it is consumed in much smaller quantities (Attia, 2018). Tables 4.6 and 4.7 give the embodied energy produced during assembly process of floors, roofs and walls.

There has been little change in material supply and selection in the last 20 years or so, so there has been little change in embodied energy consumption in dwelling per unit area (about 5 GJ/m^2). However, there is an increase in floor area per dwelling by about 20% with a corresponding increase in embodied energy (Committee, 2001; Treloar & Fay, 2005). Embodied energy represents 20–50 times the annual operational energy of most Australian buildings (Treloar & Fay, 2005). Reuse of building materials could save from 20% (glass) to 95% (aluminium) of the embodied energy (Kumar et al., 2020). Thus, in order to save on embodied energy, one can

Table 4.6 Embodied energy of assembly process of floors and roofs (Milne & Reardon, 2013)

Floors	Embodied energy (MJ/kg)	Roofs	Embodied energy (MJ/kg)
Elevated timber Floor	293	Timber frame, concrete tile, plasterboard ceiling	251
110-mm concrete slab on ground	645	Timber frame, terracotta tile, plasterboard ceiling	271
200-mm precast concrete T beam/infill	644	Timber frame, steel sheet, plasterboard ceiling	330

Table 4.7 Embodied energy of assembly process of walls (Milne & Reardon, 2013)

Wall	Embodied energy (MJ/kg)	Wall	Embodied energy (MJ/kg)
Single-skin AAC block wall	440	Single-skin AAC block wall gyprock lining	448
Single-skin stabilised (Rammed) earth wall (5% cement)	405	Steel frame, compressed fibre cement clad wall	385
Timber frame, reconstituted timber weatherboard wall	377	Timber frame, fibre cement weatherboard wall	169
Cavity clay brick wall	860	Cavity clay brick wall with plasterboard internal lining and acrylic paint finish	906
Cavity concrete block wall	465		

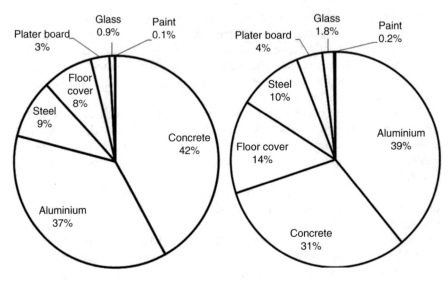

Fig. 4.6 Share of (**a**) GHG emission of production and (**b**) embodied energy of construction materials in buildings. (Biswas, 2014)

encourage smaller size dwellings (a planning issue) and the reuse of building materials (a building code issue).

In estimating the amount of embodied energy associated with a building, it is necessary to count the energy associated with the production, transportation and construction stages. Figure 4.6 shows the GHG emission of production of construction materials and percentage of total embodied energy in typical Australian buildings. It can be seen that concrete has the highest share of the GHG emission, followed by aluminium, floor cover and steel. However, in the case of embodied energy, aluminium has the highest embodied energy followed by concrete. This is because the production of aluminium requires very high amount of energy than the production of concrete (about 200 times), and GHG emission is 19 times higher than concrete production (Biswas, 2014).

In estimating the effectiveness of recycled materials, it is necessary to compare the embodied energy of the virgin and recycled materials since some recycled materials require more energy to reprocess than virgin materials. Reducing heavyweight construction systems that use masonry, and adopting lightweight construction that uses timber or light gauge steel framing to support nonstructural components, may reduce the embodied energy. If the economics of material production and delivery is not distorted by other factors, there should be a close correlation between the cost of a building material or component and its embodied energy. Thus, by trying to obtain a minimum cost solution, the designer may also achieve a most efficient embodied energy solution. It should be mentioned that the selection of materials with low embodied energy may have a negative impact on building thermal performance, which subsequently leads to more energy consumption for heating, ventilation and air conditioning (HVAC), which should actually be considered over the whole service life cycle of buildings.

4.6 Operating Energy

4.6.1 Residential End Use

Reducing operating energy consumption will have a direct effect on greenhouse gas emissions. Heating and cooling accounts for about 50% of the total energy consumption and 33% of greenhouse gas emissions from a dwelling (Memon, 2014; Ramakrishnan et al., 2015). Figure 4.7b breakdowns the projected energy consumption in 2030 by residential end use in Australia, indicating that space heating and cooling is projected about 38% of total energy consumption. Meanwhile, energy consumption by lighting and appliances is projected about 29%. The increase in appliances energy consumption may be related to the increasing adoption of new household electric goods.

The Greenhouse Account also indicated that heating and cooling accounted for 11.8% CO_2 contribution of the total from space heating and cooling in 1990, which increased to 25.3% in 2004. More usage of air conditioners also increases the energy consumption from 18% in 1990 to 42% in 2004 of the total consumption for space heating. The increase may be related to the consistent growth of population and housing; it may also be considerably affected by climate change, particularly in warmer areas. However, from 2014 to 2017, there is a decline in total residential energy consumption with the major reduction in lighting and space conditioning. This could be due to the advancement in light devices (i.e. light emitting diode globes) and the introduction of mandatory energy rating scheme (Nationwide House Energy Rating Scheme) in new residential buildings leading to high building energy efficiency.

4.6.2 Commercial End Use

In commercial buildings, Space heating, cooling and lighting are significant CO_2 contributors. According to the Greenhouse Account, HVAC accounted for 61.2% of total commercial sector energy use and 55.4% of total commercial emission, while the lighting shared 18.6% of the total energy consumption and 23.4% of the total emission from commercial end use in 2005.

Therefore, efficiency of building operating energy has become more and more important, as demonstrated by the fact that more requirements have been implemented in most building codes around the world.

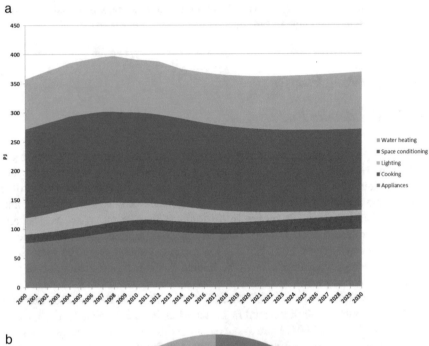

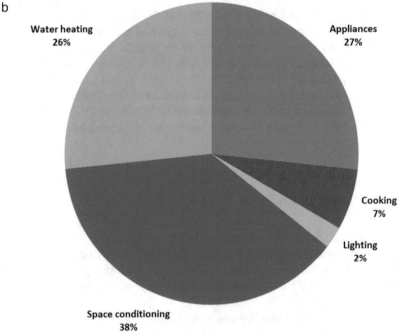

Fig. 4.7 Residential energy use in Australia (**a**) total residential end use (**b**) distribution of projected end use in 2030. (Government, 2019)

4.7 Energy Efficiency in Buildings

A building system can be considered as a complex thermodynamic system, sub-jected to internal and external thermal solicitations, with the building's envelope as the boundary of the system. The energy efficiency of buildings should include the efficiency aspects in the following elements:

- The building envelope – both conduction and solar transmission aspects
- The lighting system
- The heating and cooling system
- The ventilation system
- The hot water supply system
- The building elevators

The Japanese system also includes other factors such as:

- Direct use of natural energy (e.g. natural light and ventilation)
- Use of renewable energy (e.g. solar panel, etc.)
- Other measures for improving energy efficiency (e.g. use of unused heat, moni-toring system, operational management system, etc.)

ASHRAE GreenGuide is a useful resource for designers wishing to obtain energy-efficient designs particularly in heating, ventilation and air-conditioning areas. Currently, there are no national energy efficiency measures in the Building Codes of Australia (BCA). However, three jurisdictions, the Australian Capital Territory (ACT), South Australia and Victoria, do have measures in their BCA Appendices. Energy efficiency measure is now considered in Part 3.12 of BCA Vol.2. Other areas that the BCA might address are the use of alternative energy sources and building design for better energy efficiency.

Efficiency Principles

There are many efficiency principles to increase energy efficiency and to reduce GHG emission (Metz et al., 2007). Some of the efficiency principles are given below:

- Reduce heating, cooling and lighting load by optimisation of insulation, glazing, airtightness, thermal load shifting (for example, store energy during daytime and heating space during night through thermal energy harvest system).
- Actively utilise solar, wind and other renewable energy, create heat source or sink. Passive building design has been widely advocated to increase energy efficiency.
- Increase efficiency of appliances and HVAC equipment.
- Develop incentives and enhance awareness to change energy use behaviour.
- Apply systematic and holistic approach in design and management by consider-ing energy use over time instead of a steady time, and over all end uses in the building space instead of individual appliances.
- Consider building form, orientation, shading and related attributes.

4.8   Carbon Accounting in Construction

Carbon account is becoming an increasingly important issue when there is a growing demand from regulations and obligations in reporting greenhouse gas emission. It is not only to be applied to inform government and the Australian public as well as to meet international obligations at national and state levels, but also to be applied to make decisions in planning, design, construction and operation at an individual level.

Australian National Greenhouse and Energy Reporting Act 2007 (Lodhia, 2011) introduced a framework for the reporting, which is extracted to form the basis discussed in the following. Greenhouse gas emissions are measured in tonnes of CO_2 equivalent or CO_2-eq and divided into:

Scope 1: It concerns emission released from a facility as a direct result of the activities of the facilities, and determined by (Energy, 2017)

– fuel combustion
– fugitive emission: mainly released from the extraction, production, processing and distribution of fossil fuels
– industrial process emission: released from the consumption of carbonates and the use of fuels as feedstocks or as carbon reductants
– waste emission: released from landfill and wastewater treatment.

Scope 2: It concerns emissions in a form of indirect emission that occurs principally at electricity generators as a result of electricity consumption at another facility.

In general, there are four methods to measure emissions (Energy, 2017):

1. Using national greenhouse accounts, which is the default method used by Australian Department of Climate Change and Energy Efficiency and is within the international guidelines adopted by the United Nations Framework Convention on Climate Change for the estimation of greenhouse gas.
2. A facility-specific method, using industrial sampling and Australian or international standards or equivalent for the analysis of fuels and raw materials to provide more accurate estimation of emissions.
3. A facility-specific method, using Australian and international standards or equivalent for both sampling and analysis of fuels and raw materials.
4. Direct monitoring of emission systems, either continuous or a periodic basis.

For sustainable buildings and construction, Method 1 will be discussed. Fundamentally, it specifies the use of designed emission factors, for example, in terms of kilogram of CO_2-eq per kilojoules of energy use, in the estimation of emissions. The emission factors are national average factor determined by Australian Department of Climate Change and Energy Efficiency using Australian Greenhouse Emissions Information Systems. For Scope 2, emission factors are updated annually to reflect the latest information on the mix of electricity generation sources. The more use of renewable energy leads to low emission factors. The following gives the details of estimation of emissions based on Method 1.

4.8.1 Emissions from Fuel Combustion

Emissions from fuel combustion is estimated by,

$$E_{ij} = \frac{Q_i \times EC_i \times EF_{ij}}{1000} CO_2 - e \text{ tonnes}$$

where E_{ij} is the emissions of gas type (j) from each fuel type (i) in CO_2-eq tonnes, Q_i is the quantity of fuel type (i) combusted in operation of the facility during the year, EC_i is the energy content factor of fuel type (i), and EF_{ij} is the emission factor for emission gas type (j) released from the combustion of fuel (i).

4.8.2 Emissions from Electricity

Scope 2 emission or emission from electricity is estimated by:

$$Y = Q \times EF / 1000$$

where Y is the Scope 2 emissions measured in CO_2-eq tonnes. Q is the quantity of electricity purchased from the electricity grid during the year and consumed from the operation of the facility measured in kilowatt hours (1 GJ = 278 kWh). EF is the Scope 2 emission factor, in kilograms of CO_2-eq emissions per kilowatt hour.

The Scope 2 emission factors are assigned as state-based emission scenario because the electricity production methods and production efficiency differ in each state of Australia. The emission factors also consider the interstate electricity flows between states. Table 4.8 shows the emission factor for the financial year of 2018/2019.

Table 4.8 Emission factors for electricity in Australia (Government, 2017)

	Emission factor kg CO_2-eq/kWh
New South Wales and Australian Capital Territory	0.83
Victoria	1.08
Queensland	0.79
South Australia	0.49
South West Interconnected System in Western Australia	0.70
North Western Interconnected System in Western Australia	0.62
Darwin Katherine Interconnected System in the Northern Territory	0.59
Tasmania	0.14
Northern Territory	0.64

4.9   Implications of Climate Change to Residential Building Energy

While the residential building sector may contribute significantly to the GHG emissions and climate change, the changing climate also affects the building energy demand of existing buildings. Buildings are expected to operate for several decades and the design and operation of buildings for changing climate should be considered. Data from the Australian housing survey by the Australian Bureau of Statistics shows that about 45% of the current housing stock is between 20 to 50 years of age and 18% older than 50 years (Australia, 2008). Over such a long lifespan, the local climate conditions, such as the average ambient temperature and relative humidity, may undergo considerable changes and thus impacting on the heating and cooling energy requirement of residential houses (Yau & Hasbi, 2013; Zhai & Helman, 2019).

It was shown that regional electricity demand trend is closely correlated with heating degree days (HDD) as well as cooling degree days (CDD), based on the analysis by the National Electricity Management Marketing Company in Australia (Ahmed et al., 2012). Heating degree days (HDD) and cooling degree days (CDD) are used as indicators for analysing weather-related energy consumption in buildings. Here, Degree days is defined as "a summation of the differences between a reference temperature and the outdoor air temperature over a period of time. The reference temperature is known as base temperature and is defined as the outdoor temperature at which the heating (or cooling) systems are not required to maintain comfort conditions" (ASHRAE, 2013; Lee et al., 2014). When the outdoor temperature is below (above) the base temperature, the heating (cooling) systems are required to operate, and therefore, deviations result in increased energy requirements.

It was also predicted that climate change may considerably impact the electricity demand, especially the peak demand (Isaac & Van Vuuren, 2009; Wang et al., 2010). Studies carried out in the UK (Jeswani et al., 2008) and New Zealand (Camilleri et al., 2001) found that a mean temperature rise might lead to energy saving and GHG emission reduction. Similarly, simulations for the United States and Switzerland indicated that a warming climate could reduce the heating energy requirement in relatively cold climates (Christenson et al., 2006; Wang & Chen, 2014). On the other hand, the warming climate will result in an increase in the cooling energy requirement, which may eventually offset or exceed the benefit from the heating energy saving (Frank, 2005; Wang et al., 2010).

4.10   Summary

Energy and carbon emissions are the commonly used metrics to quantitatively assess the sustainable development in built environment. Buildings and infrastructure are expected to last 50–200 years, and the efficient use of energy as well as the emission reduction during the life cycle is crucial to achieve sustainable development. This chapter initially discusses various sources of energy, their production and consumption with the information on the global energy consumption trend. A special attention is given to construction sector with the introduction to embodied energy and operating energy followed by the carbon-accounting methods for these energy types. The preliminary information of different energy efficiency measures to reduce embodied energy and operating energy is also presented. The detailed discussion on energy efficiency measures is presented in Chap. 7.

References

Ahmed, T., Muttaqi, K., & Agalgaonkar, A. (2012). Climate change impacts on electricity demand in the State of New South Wales, Australia. *Applied Energy, 98*, 376–383.

ASHRAE, A. (2013). *Standard 55–2013: "Thermal environmental conditions for human occupancy"*. ASHRAE. In *Atlanta USA*.

Attia, S. (2018). Net Zero Energy Buildings (NZEB): Concepts, frameworks and roadmap for project analysis and implementation: Butterworth-Heinemann.

Australia, Y. B. (2008). *Australian bureau of statistics*. Canberra, Australia, 161.

Biswas, W. K. (2014). Carbon footprint and embodied energy consumption assessment of building construction works in Western Australia. *International Journal of Sustainable Built Environment, 3*(2), 179–186. https://doi.org/10.1016/j.ijsbe.2014.11.004

Camilleri, M., Jaques, R., & Isaacs, N. (2001). Impacts of climate change on building performance in New Zealand. *Building Research & Information, 29*(6), 440–450.

Chau, C. K., Leung, T. M., & Ng, W. Y. (2015). A review on Life Cycle Assessment, Life Cycle Energy Assessment and Life Cycle Carbon Emissions Assessment on buildings. *Applied Energy, 143*, 395–413. https://doi.org/10.1016/j.apenergy.2015.01.023

Christenson, M., Manz, H., & Gyalistras, D. (2006). Climate warming impact on degree-days and building energy demand in Switzerland. *Energy Conversion and Management, 47*(6), 671–686.

Committee, A. S. o. t. E. (2001). Independent Report to the Commonwealth Minister for the Environment and Heritage. In: CSIRO Publishing on behalf of the Department of the Environment and Heritage ….

Dudley, B. (2018). BP statistical review of world energy. *BP Statistical Review*, London, UK. Accessed 6 Aug 2018.

Energy, T. D. o. t. E. a. (2017). *Technical Guidelines for the estimation of emissions by facilities in Australia*. Retrieved from http://www.environment.gov.au

Frank, T. (2005). Climate change impacts on building heating and cooling energy demand in Switzerland. *Energy and Buildings, 37*(11), 1175–1185.

Government, A. (2017). *National greenhouse accounts factors*. Department of Environment.

Government, A. (2019). *Australian energy update 2019*. Commonwealth of Australia.

Grübler, A., & Fisk, D. J. (2013). *Energizing sustainable cities: Assessing urban energy*. Routledge.

IEA, I. (2019). *World energy statistics and balances, IEA*. OECD.

Isaac, M., & Van Vuuren, D. P. (2009). Modeling global residential sector energy demand for heating and air conditioning in the context of climate change. *Energy Policy, 37*(2), 507–521.

Jeswani, H. K., Wehrmeyer, W., & Mulugetta, Y. (2008). How warm is the corporate response to climate change? Evidence from Pakistan and the UK. *Business Strategy and the Environment, 17*(1), 46–60.

Kumar, D., Alam, M., Zou, P. X., Sanjayan, J. G., & Memon, R. A. (2020). Comparative analysis of building insulation material properties and performance. *Renewable and Sustainable Energy Reviews, 131*, 110038.

Lee, K., Baek, H.-J., & Cho, C. (2014). The estimation of base temperature for heating and cooling degree-days for South Korea. *Journal of Applied Meteorology and Climatology, 53*(2), 300–309. https://doi.org/10.1175/jamc-d-13-0220.1

Lodhia, S. (2011). The Australian National Greenhouse and Energy Reporting Act and its implications for accounting practice and research. *Journal of Accounting & Organizational Change.*

Memon, S. A. (2014). Phase change materials integrated in building walls: A state of the art review. *Renewable and Sustainable Energy Reviews, 31*(0), 870–906. https://doi.org/10.1016/j.rser.2013.12.042

Metz, B., Davidson, O., Bosch, P., Dave, R., & Meyer, L. (2007). *Climate change 2007: Mitigation of climate change.* Cambridge University Press.

Milne, G., & Reardon, C. (2013). *Materials, embodied energy–Your home Australia's guide to environmentally sustainable homes.* Commonwealth of Australia (Department of Industry).

Ramakrishnan, S., Sanjayan, J., Wang, X., Alam, M., & Wilson, J. (2015). A novel paraffin/expanded perlite composite phase change material for prevention of PCM leakage in cementitious composites. *Applied Energy, 157*, 85–94. https://doi.org/10.1016/j.apenergy.2015.08.019

Solomon, S., Manning, M., Marquis, M., & Qin, D. (2007). *Climate change 2007-the physical science basis: Working group I contribution to the fourth assessment report of the IPCC* (Vol. 4). Cambridge University Press.

Treloar, G., & Fay, R. (2005). Building materials selection–Greenhouse strategies. *Environment Design Guide*, 1–9.

Wang, H., & Chen, Q. (2014). Impact of climate change heating and cooling energy use in buildings in the United States. *Energy and Buildings, 82*, 428–436.

Wang, X., Chen, D., & Ren, Z. (2010). Assessment of climate change impact on residential building heating and cooling energy requirement in Australia. *Building and Environment, 45*(7), 1663–1682.

Yau, Y., & Hasbi, S. (2013). A review of climate change impacts on commercial buildings and their technical services in the tropics. *Renewable and Sustainable Energy Reviews, 18*, 430–441.

Zhai, Z. J., & Helman, J. M. (2019). Implications of climate changes to building nergy and design. *Sustainable Cities and Society, 44*, 511–519.

Chapter 5
Materials and Water

5.1 Introduction

This chapter discusses the use of resources such as materials and water in the construction of buildings as well as its implications to the environment. All products used in building construction require transforming raw materials extracted from the environment. There are various informed strategies for using material inputs efficiently, such as using recycled and reclaimed wastes as inputs, will be discussed. Lack of water is another major constraint to industrial and economic growth. The concepts of embodied energy and embodied water are introduced in this chapter. The ways to improve water efficiency in buildings will also be discussed.

5.2 Sustainable Resource Management

The resources used in the construction section can be viewed under six major themes as follows:

- Sustainable sites – Select the site to maximize the use of existing infrastructure by selecting the sites close to already-developed areas and public transport, protect or restore habitat, minimize impact of construction on storm water, reduce heat island effects, etc.
- Materials and resources – Reuse of buildings after renovation, recycle of materials, use of local materials, use of rapidly renewable materials, etc.
- Water efficiency – Minimize water usage for landscaping, reduce water usage in building with efficient fixtures and rainwater harvesting and use innovative wastewater technologies.
- Waste management – Reduce the amounts of waste generated during all stages of construction process. Reuse, recycle and recover as much waste as possible. Adopt waste heat recovery, practice controlled dumping, etc.

© Springer Nature Switzerland AG 2021
X. Wang, S. Ramakrishnan, *Environmental Sustainability in Building Design and Construction*, https://doi.org/10.1007/978-3-030-76231-5_5

- Energy and atmosphere – Proper commissioning of building's energy system, minimization of energy usage, on-site renewable energy and proper usage of refrigerants.
- Indoor environmental quality – Minimize sick building syndrome, improve indoor air quality, minimize the impact of construction activities on indoor air, improved daylighting, etc.

In this chapter, we will discuss the sustainable resource management of materials and water. The following chapters will discuss other aspects.

5.2.1   Global Material Consumption Trend

It has been reported that globally, the total amount of materials extracted, harvested and consumed is about 62 billion metric tonnes (Gt) in 2008, showing an increase of 60% since 1980 (Lutter et al., 2014). The primary sectors for increased global demand are construction minerals, biomass for food and feed and fossil energy carriers. These three groups account for 80% of total global material extraction. Among the different resource extraction, biomass extraction is about 33%, followed by construction minerals, fossil energy carriers and metal ores being 30%, 20% and 13%, respectively. It is also truth that the resource extraction from nonrenewable sources has increased during the last century, while showing the decline in the extraction from renewable sources. Figure 5.1 shows the material extraction rates from 1980 to 2008, representing the contribution of each sector.

The construction sector represents the major contribution to the material consumption in many countries around the world, including Australia. In Australia, the percentage of consumption in construction sector and other sectors is summarised, as seen in Table 5.1. It is understood that approximately, 90% of the production of construction materials was consumed by construction.

5.2.2   Sustainable Material Management

As discussed before, any product is embodied with energy, which is consumed during processes in association with the acquisition of natural resources, production and supply to final consumption. The processes may be involved with mining, manufacturing, transport and other sectors. Therefore, proper selection of materials for construction may not only lead to the significant improvement of indoor comfort, but also reduce environmental impacts, which is also affected by the effectiveness in reuse/recycle and transportation of materials and their products.

To reduce the environmental impact due to material consumption, two approaches may be considered through the life cycle of building construction and service. One

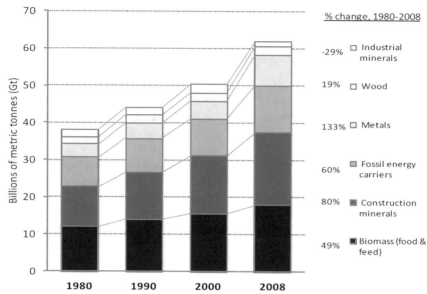

Source: SERI (Sustainable Europe Resource Institute) material flows database.

Fig. 5.1 Global material extraction trend. (Lutter et al., 2014)

Table 5.1 Consumption of materials by end use in construction (Newton, 2001)

Materials	% of total consumption
Iron and steel	12
Construction materials (concrete, brick, gravel, sand)	90
Timber products[a]	55
Plastic products	27

[a]In selecting timber for use, it is important to distinguish plantation timber (which is a renewable material) and timber from old growth forest (which is much less renewable).

approach is to directly reduce new material consumption, and another is to reduce embodied energy (and embodied water).

Reduce material consumption by:

- Reducing the consumption of materials directly during the planning and design stage.
- Using recycled materials and products with recycled materials. Recycling building materials can reduce embodied energy substantially. For example, aluminium is 100% recyclable. Recycling aluminium reduces embodied energy by 95%, while recycling steel reduces embodied energy by 72% (Milford et al., 2011).
- Reuse materials and products.

- Using durable materials and products, with their service life equivalent to projected building service life.
- Consuming rapidly renewable materials.
- Using innovative and emerging construction methods to reduce material consumption.

Figure 5.2 shows an aspect of reducing the material consumption at design stage by eliminating the exterior and interior plastering in brick veneer buildings. Although the brick surfaces are exposed visibly, they do not reduce the aesthetic appearance of the building.

Construction methods also significantly contribute to the amount of materials consumed. For instance, on-site construction without proper planning would make up enormous amounts of offcuts and waste, and cutting down such offcuts with proper planning would reduce the amount of material consumption. Bricks when cut to half or any other size will result in the remaining half unused. Similarly, reinforcement is heavily wasted during the construction as the offcuts are generally not used. In order to minimise the material consumption, we would like to discuss two innovative or emerging construction methods that could significantly reduce material consumptions and, hence, waste:

1. Off-site or modular construction: Modular construction is an innovative construction process whereby a building is constructed off-site (possibly in a factory), under controlled and well-planned conditions, using the same materials and meeting the standards to conventionally constructed buildings. As this construction method produces similar modular units again and again, the material consumption is well planned ahead without significant offcuts. Most importantly, they use permanent formworks for construction that could substantially eliminate the material consumptions for formwork in on-site construction. Figure 5.3 shows the construction of a modular unit and assembly on-site.

Fig. 5.2 Brick exterior walls without plaster (Green & LEED, 2010)

Fig. 5.3 Modular construction process (Ingram, 2017)

Fig. 5.4 3D printing of concrete buildings (ALL3DP., 2020)

2. Freeform or digital construction method: The freeform construction methods construct building on-site without the use of formwork for construction. The geometry of the building is digitally embedded into the robotics, which moves in the designated path for constructing the building as layer-by-layer. It is similar to three-dimensional (3D) printing of plastic objects; however, it uses concrete as a key construction material with or without reinforcement. As this method freely constructs the building without the formwork, the material consumption for formworks could be significantly reduced, in addition to reducing the construction material waste. This method is in its infant stage and requires a significant research and innovation to implement in construction (Fig. 5.4).

Reduce embodied energy by:

- Using materials that require less energy to extract and process.
- Using materials and products, which may need less transportation for building construction and service (i.e. locally sourced material).
- Recycling building materials can reduce embodied energy substantially. For example, aluminium is 100% recyclable. Recycling aluminium reduces embodied energy by 95%, while recycling steel reduces embodied energy by 72%.

Figure 5.5 is an example of using locally sourced soil for making a cascade system with cement-stabilised rammed earth (CSRE) walls. CSRE uses soil as a main ingredient with very little amount cement for stabilisation. The use of excavated soil

Fig. 5.5 Cement-stabilised rammed earth (CSRE) wall construction (Green & LEED, 2010)

for constructing the wall will not only reduce the landfill soil waste, but also reduces the material consumption for construction.

Meanwhile, the process to extract materials by mining or harvesting for building and construction can cause adverse impacts on landscape and biodiversity, leading to land depletion or degradation and loss of natural species. The process with less environmental impacts should also be considered in the selection of materials. However, in addition to costs, material section should be balanced with the building performance requirements, for example, thermal performance, which may effectively make a significant difference in operating energy consumption of buildings.

5.3   Sustainable Water Management

5.3.1   Overview

Water is an essential part of every life. Human consumes water for domestic use (their day-to-day activities), agricultural use and industrial demands. Globally, the total volume of water on the earth is about 1400 million km³ of which only 2.5%, or about 35 million km³, is freshwater. About two-thirds of the freshwater remains as ice caps and glaciers, which human cannot use. The remaining one-third of the freshwater is accessible by human as either groundwater or surface water. Fig. 5.6 shows the distribution of the earth's water.

On average, about 312 billion gallons of surface water and 77 billion gallons of groundwater are consumed by human daily. Like any other resources, the water consumption is also continuously increasing, and the major factors causing increasing water demand are increasing population growth, industrial development and the expansion of irrigated agriculture. Agriculture accounts for more than 70% of freshwater drawn from lakes, rivers and underground sources.

In Australia, 70% of total water withdrawal is used for agriculture (Calzadilla et al., 2010). Rainfall is the major source of water in Australia, becoming

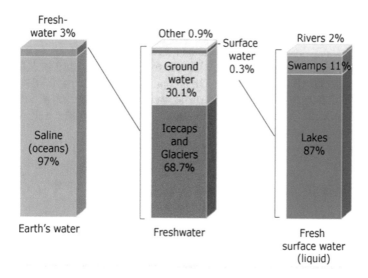

Fig. 5.6 Distribution of the earth's water. (Shiklomanov, 1993)

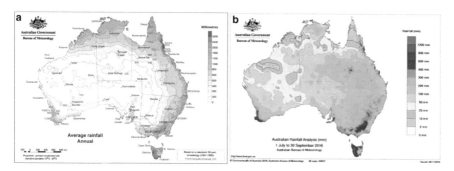

Fig. 5.7 Australian rainfall analysis (**a**) 30 year average from 1961–1990 (**b**) 1 July to 30 September 2016. (Australian Bureau of Meteorology 2016)

- run-off that goes into rivers, lakes, dams and aquifers
- groundwater recharge by infiltration that varies the water table
- water storage on the land surface or underground to supply agriculture, industry and urban users, including large dams, farm dams and aquifers

The quantity of water stored above ground, such as dams, and below ground, such as aquifers, is determined by the volumes of run-off and recharge from rainfall. However, in Australia, there has been an extreme variability in rainfall both across the continent and from year-to-year. As a matter of fact, there are considerable areas in Australia that have been experiencing prolonged draught, as rainfall fell short of the average quite significantly in the past years, as shown in Fig. 5.7. Water saving now becomes one of the most important issues in Australia.

The total water extraction in Australia was 76,159 GL in 2016–2017 (17.5% lower than 2013–2014). The electricity and gas supply industry extracted 59,869 GL mainly for hydroelectricity generation and was discharged back to the environment without consumption. The total water consumption was 16,558 GL. Households consumed 1909 GL. Total water consumption increased by 23.6% between 2008–2009 and 2016–2017. The industrial sector has an increase of 26.8% since 2008–2009, while household consumption increased by only 1.9% over the same period.

5.3.2   Water Efficiency and Recycling

Water movement between biosphere, hydrosphere, atmosphere and lithosphere is known as a biogeochemical cycle. Figure 5.8a shows the movement of water between different elements in the earth. Nevertheless, the water movement in the urban environment is the crucial part that requires assessment, because urban environment may disturb the natural water cycle due to the urban building density, less vegetation and impervious surfaces. Water cycle in urban environment can be described by Fig. 5.8b, starting from dams and water tanks for some households, sourced from rainfall. The water would then be treated and delivered to household. Household water may also be sourced from recycled water. It has been found that in Australia, households use about 59% of urban water; of that, 54% of the water used in the average Australian household is used for flushing toilets and watering gardens.

The water use efficiency should be improved to prevent continuously increasing water extraction to meet the demands of increasing population and the human lifestyle. The following are few ways to increase water efficiency:

- Changing human activities (shifting to more water-efficient crops, changing industrial processes away from water-intensive production)
- Adopting existing technology (such as drip irrigation, low-flow toilets and better industrial processes)
- Adopting water efficient irrigation technology
- Changing the water consumption practices such as preventing irrigating during the day, preventing the use of potable water for landscape irrigation
- Identifying and preventing wasteful leaks
- Charging proper prices for water

On the other hand, when assessing efficiency of water usage in buildings, the following factors should be considered:

- Water-saving devices: low-flow fixtures (toilets, urinals, faucets and shower-heads), no-flow fixtures (composting toilets and waterless urinals) and controls (infrared sensors)
- Use of rainwater: collection, storage and, if necessary, treatment of rain falling on structures in the built environment

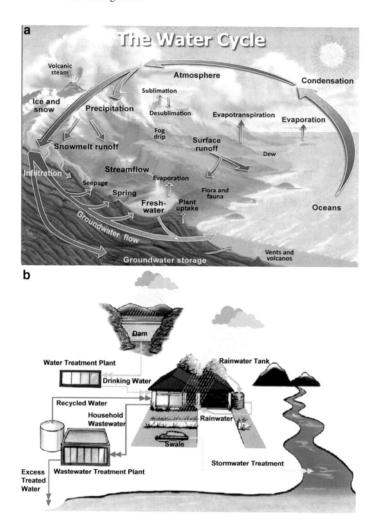

Fig. 5.8 (a) Global water cycle (Fandel et al., 2018) (b) urban hydrological cycle. (Dimitriadis, 2005)

- Use of storm water: collection, storage and treatment of rain falling on hard surfaces and running off the built environment
- Reuse or recycling of grey water: reuse domestic sewage effluents or municipal wastewater, preferably from which industrial effluents containing processing chemicals have been segregated. The recycled water may be reclaimed from bathroom and laundry effluents (grey water) or from the entire domestic sewage stream (black water) or municipal wastewater.

A study on the efficiency of showerheads and washer on the domestic water consumption in 150 Australian households is shown in Figs. 5.9 and 5.10 (Willis et al., 2013). The showerhead efficiency was categorised into low, medium and

Description	Showerhead efficiency clusters		
Efficiency category	Low	Medium	High
Weight range	$w \leq 2$	$2 < w < 4$	$w \geq 4$
No. of households in cluster ($n = 151$) Proportion (%)	50 (33.1%)	42 (27.8%)	59 (39.1%)
No. of people in cluster ($n = 495$) Proportion (%)	190 (38.4%)	124 (25%)	181 (36.6%)
Per capita shower consumption per day (L/p/d)	64.7	46.8	33.6
Household shower consumption per day (L/hh/d)	245.7	138.1	103.1
Per capita shower consumption per annum (kL/p/a)	23.6	17.1	12.3
Household shower consumption per annum (kL/hh/a)	89.7	50.5	37.6

Fig. 5.9 Showerhead efficiency assessment. (Willis et al., 2013)

Description	Clothes washer efficiency clusters		
Efficiency category	Low	Medium	High
Star rating range	1–2.5	3–3.5	4–6
Category (L/wash)	120–170	80–119	40–79
No. of households in cluster ($n = 148$) Proportion (%)	38 (25.7%)	40 (27.0%)	70 (47.3%)
No. of people in cluster ($n = 486$) Proportion (%)	148 (30.5%)	119 (24.5%)	219 (45.0%)
Per capita clothes washer consumption (L/p/d)	53.0	36.3	14.4
Household clothes washer consumption (L/hh/d)	206.4	108.0	45.2
Per capita clothes washer usage per annum (kL/p/a)	19.3	13.3	5.3
Household clothes washer usage per annum (kL/hh/a)	75.4	39.4	16.5

Fig. 5.10 Clothes washer efficiency assessment. (Willis et al., 2013)

high. It can be seen that the household shower consumption can be reduced from 89.7 kl/hh/a to 37.6 kl/hh/a when the low-efficiency showerheads are replaced by high-efficiency showerheads. Further analysis on economic assessments showed that the showerhead upgrade would save 69 AUD/year on water consumption with the payback period of 2 years (this is without considering the cost-savings for water heating; such consideration would further reduce the payback period). On the other hand, upgrading the domestic washers to high-efficiency washers would save 86 AUD/year with the payback period of 7 years.

5.3.3   *Water Quality and Treatment*

Another important issue with water resources is the contamination of the water supplies. Figure 5.11 shows the contaminants in water on short and long terms. The discharge of hazardous waste products particularly in manufacturing processes could be a major cause of contamination in the long run. The quality degradation of natural water has serious consequences on human health as reported in Table 5.2. It must be stressed that, apart from natural contaminations, the anthropogenic pollution has major concerns that are irreversible.

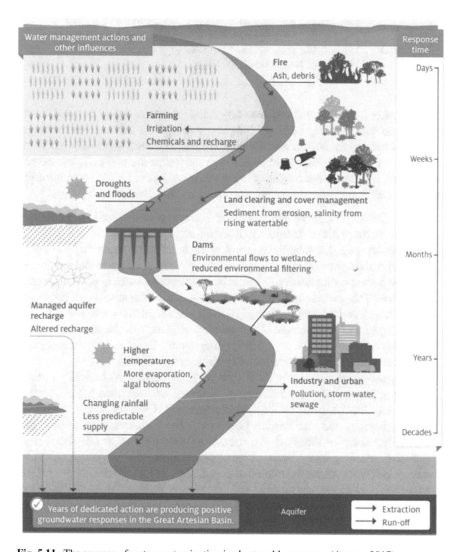

Fig. 5.11 The sources of water contamination in short and long terms. (Argent, 2017)

Table 5.2 Problems, causes and concerns of water contamination (Foster et al., 1998)

Problem	Causes	Concerns

Anthropogenic pollution	Inadequate protection of vulnerable aquifers against human-made discharges and leachates from:
	  Urban and industrial activities
	  Intensification of agricultural cultivation
		Pathogens, nitrates, ammonium salts, chlorine, sulphates, boron, heavy metals, dissolved organic carbon, aromatic and halogenated hydrocarbons
		Nitrates, chlorine, pesticides

Naturally occurring contamination	Related to pH-Eh evolution of groundwater and dissolution of minerals (aggravated by anthropogenic pollution and/or uncontrolled exploitation)	Mainly iron, fluorine and sometimes arsenic, iodine, manganese, aluminium, magnesium, sulphates, selenium and nitrates
Well-head contamination	Inadequate well design and construction allowing direct intrusion of polluted surface water or shallow groundwater	Mainly pathogens

Wastewater treatment also becomes very critical as some lightly polluted wastewater can be used for many noningestion purposes. With the advanced technologies, they could be used for human consumption as well. The components of wastewater treatment are generally described as the preliminary, primary, secondary and tertiary stages, with advanced treatment being added to the third stage when the effluent is to be returned to potable or near-potable standard (Radcliffe, 2004).

- *The preliminary stage* involves physical screening of the arriving influent to remove coarse particles such as small stones, sand and gravel.
- *The primary stage* involves the removal of the most of the remaining particulate matter, involving comminution if necessary, followed by coagulation and/or flocculation before sedimentation. Alternatively, filtration can be applied after sedimentation. This processing will remove about half of the suspended solids and reduce the biological oxygen demand (BOD) and will also remove about 10% of the nitrogen and phosphorus. If the primary effluent is to be discharged, it is likely to be disinfected. However, processing to only the primary stage is not now perceived adequate to meet current-day discharge standards.
- *The secondary stage* usually involves a range of aerobic biological processes aiming at microbially metabolising the dissolved or suspended organic matter. These may involve using either slow rate suspended growth processes such as aerated lagoons and stabilisation ponds or faster processes such as activated sludge technologies. Alternatively, fixed film processes may be adopted such as the use of trickling filters. Some of the organic matter in the wastewater provides energy and nutrients for the microbial populations, the remainder being oxidised to carbon dioxide, water and other end products. A secondary sedimentation follows to remove the biomass produced. The biomass may then be treated by aerobic or anaerobic digestion, by composting or by other technologies. These processes remove up to half of the nitrogen and convert the phosphorus to phosphates. There may be a further filtration of the effluent stream, which is then

disinfected. About 80–95% of the BOD and suspended solids are removed in the
secondary treatment.

- *Tertiary treatment* is usually the minimum for discharge to water bodies, particularly inland. Its processes involve further removal of colloidal and suspended
solids by chemical coagulation and filtration, with the removal of specific metals,
pathogens and nutrients. Activated carbon can be used to adsorb hydrophobic
organic compounds, and lime can precipitate various cations and metals at a
high pH.

5.3.4 Rainwater Tanks

Rainwater tanks offer a smaller environmental footprint than dams or desalination
plants at a domestic or an industrial level. The main advantage of having a rainwater
tank is that the owners could avoid the consequences of city-wide water restrictions;
they may partially offset their annual water bill and, in some areas, rainwater may
offer a better aesthetic than the city drinking water supply.

The yield of a rain water tank is directly related to the rainfall and the tank size.
Although day to day fluctuations in rain fall is significant, the yield of a tank over a
period is less volatile than large water storage options such as dams. The rain water
tank yield primarily depends on:

- rainwater collection area (roof size)
- tank size
- the number of occupants in the house (and therefore usage)
- garden requirements
- whether the tank is plumbed into the house, and if so, to which areas

Considering above factors, a simple formula to determine the rain water tank
yield is shown in the following equation:

$$\text{Yield} = A \times (\text{Rainfall} - B) \times \text{Roof Area}$$

where A is the efficiency of collection and a value of 0.8 is generally considered.
B is the loss associated with adsorption and wetting of surfaces and a value of 2 mm
per month (24 mm per year) is used. However, it must be noted that this is a simple
model to calculate instantaneous rain water storage, and the daily models are used
to assess the rain water tank storage at any time. The flow chart is shown in Fig. 5.12
is the estimation of rain water tank yield using the daily model. The storage of the
tank at any time depends on many factors including initial water storage, daily rainfall and daily consumption.

The effects of different factors on the rain water tank yield are shown in Fig. 5.13.
It can be seen that the catchment area (or roof area in buildings), tank size and
annual rainfall play a major role in determining the rain water tank yield. Factors
like rainfall pattern and climate scenarios have less influence on the yield.

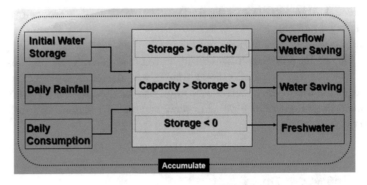

Fig. 5.12 Rain water tank yield estimation using the daily model

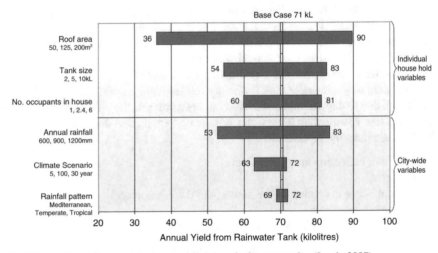

Fig. 5.13 Factors influencing the storage capacity of rainwater tanks. (Jacob, 2007)

5.3.5   Desalination, Recycling and Energy

Desalination and recycling of water plants have been seen as an effective means to cope with water shortage in Australia. Desalination in Australia is a relatively expensive way to produce drinking water. However, over the past decade, improved technologies have reduced the cost of desalination from between $3.00 and $5.00 a cubic metre ($m^3$) to between $0.46 and $0.80/m^3$, depending on local conditions (Pink, 2010). It should be reminded that both desalination and recycling process consume energy. It should also be carefully considered that energy use for water treatment has been associated with environmental costs, such as greenhouse gas emissions. The process also produces a brine waste product that has potentially hazardous effects on the marine environment.

Table 5.3 describes the energy consumption related to water treatment, reclamation and desalination. The water extraction from different sources in major Australian urban cities is shown in Fig. 5.14. It can be observed that South East Queensland,

Perth and Adelaide have already started using desalination water for the consumption due to the limited availability of natural water sources. Compared with 2014/2015, during 2015/2016 Adelaide reported a 66% decrease in desalinated water supplemented by an 18% increase in surface water.

5.4 Summary

The resource efficiency is a major concern in built environment as the current practices use more nonrenewable resources than the renewable. This chapter discussed sustainable resource management with the discussion on resource efficiency in materials and water. The global material consumption trend showed the necessity of implementation principles of efficient material consumption for the construction sector. Various methods of sustainable material management methods and strategies are discussed. This will be benefit to the engineers, teachers and students to transform the current practices with the focus on efficient material consumption. The water efficiency in built environment is also discussed with the methods to the efficient water consumption in buildings and the consequences of unsustainable water consumption practices. Finally, different supply and demand management technologies for water consumption in built environment are presented.

Table 5.3 Energy consumption for various water treatment options (Australia, 2008)

Water supply option	Energy use in kWh per kl
Reverse osmosis of sea water (sea water desalination)	3.0–5.0
Brackish reverse osmosis (still water desalination)	0.7–1.2
Municipal wastewater reclamation	0.8–1.0
Conventional water treatment	0.4–0.6

Major urban centre	Surface water (W1)		Groundwater (W2)		Desalination (W3.1)		Recycled water (W4)		Total	
	2014–15	2015–16	2014–15	2015–16	2014–15	2015–16	2014–15	2015–16	2014–15	2015–16
Sydney	516,041	534,642	0	0	0	0	38,280	38,465	554,321	573,107
Melbourne	401,899	432,886	0	0	0	0	13,059	16,717	414,958	449,603
South East Queensland	284,202	289,524	9,218	8,730	1,161	1,524	16,259	16,739	310,840	316,517
Perth	47,519	20,100	122,127	136,879	119,457	138,645	7,564	8,633	296,667	304,257
Adelaide	122,634	144,346	0	0	22,725	7,686	5,054	4,373	150,413	156,405
Canberra	47,114	50,403	0	0	0	0	4,352	4,056	51,466	54,459
Darwin	40,530	38,034	5,139	5,758	0	0	492	80	46,161	43,872

Fig. 5.14 Contribution of different water sources for major cities in Australia. (Meteorology, 2017)

References

ALL3DP. (2020). 3D Printed House: 20 Most Important Projects [Online]. Available: https:// all3dp.com/2/3d-printed-house-3d-printed-building/ [Accessed].

Argent, R. (2017). Australia state of the environment 2016: inland water, independent report to the Australian Government Minister for the Environment and Energy. *Commonwealth of Australia 2017 Australia state of the environment 2016: inland water is licensed by the Commonwealth of Australia for use under a Creative Commons Attribution 4.0 International licence with the exception of the Coat of Arms of the Commonwealth of Australia, the logo of the agency responsible for publishing the report and some content supplied by third parties. For licence conditions see creative commons. org/licenses/by/4.0. The Commonwealth of Australia has made all reasonable efforts to identify and attribute content supplied by third parties that is not licensed for use under Creative Commons Attribution 4.0 International., 10*, 94.

Australia, Y. B. (2008). Australian bureau of statistics. Canberra, Australia, 161.

Calzadilla, A., Rehdanz, K., & Tol, R. S. (2010). The economic impact of more sustainable water use in agriculture: A computable general equilibrium analysis. *Journal of Hydrology, 384*(3–4), 292–305.

Green and LEED (2010). MTR Jayasinghe.

Dimitriadis, S. (2005). *Issues encountered in advancing Australia's water recycling schemes.* Department of Parliamentary Services.

Fandel, C. A., Breshears, D. D., & McMahon, E. E. (2018). Implicit assumptions of conceptual diagrams in environmental science and best practices for their illustration. *Ecosphere, 9*(1), e02072.

Foster, S., Lawrence, A., & Morris, B. (1998). *Groundwater in urban development: Assessing management needs and formulating policy strategies.* The World Bank.

Ingram, k. (2017). Modular and offsite construction: evolution or revolution? [Online]. Available: https://blog.ifs.com/2017/12/modular-and-offsite-construction-evolution-or-revolution/ [Accessed].

Jacob, M. (2007). *The economics of rainwater tanks and alternative water supply options. A report prepared for Australian Conservation Foundation.* Nature Conservation Council (NSW) and Environment Victoria.

Lutter, S., Giljum, S., & Lieber, M. (2014). *Global material flow database.* Vienna University of Economics and Business (WU).

Meteorology, B. o. (2017). *National performance report 2015–16: Urban water utilities.* Retrieved from Melbourne.

Milford, R. L., Allwood, J. M., & Cullen, J. M. (2011). Assessing the potential of yield improvements, through process scrap reduction, for energy and CO2 abatement in the steel and aluminium sectors. *Resources, Conservation and Recycling, 55*(12), 1185–1195. https://doi.org/10.1016/j.resconrec.2011.05.021

Newton, P. W. (2001). Human settlements-Australia state of the environment report 2001 (Theme report).

Pink, B. (2010). *Australia's environment issues and trends 2010.* Australian Bureau of Statistics.

Radcliffe, J. C. (2004). *Water recycling in Australia: A review undertaken by the Australian academy of technological sciences and engineering.* Australian Academy of Technological Sciences and Engineering.

Shiklomanov, I. (1993). *World fresh water resources, water in crisis: A guide to the world's fresh water resources.* Oxford University Press.

Willis, R. M., Stewart, R. A., Giurco, D. P., Talebpour, M. R., & Mousavinejad, A. (2013). End use water consumption in households: Impact of socio-demographic factors and efficient devices. *Journal of Cleaner Production, 60*, 107–115.

Chapter 6
Sustainable Waste Management

6.1 Introduction

Traditional approach to waste management relies on the natural environment to absorb and assimilate unwanted by-products. Environmental impacts associated with waste disposal include land contamination, methane emission, odour, toxicity and consumption of land resources. Different types of waste, disposal methods, associated environmental impacts and waste management aspects are discussed in this chapter. Furthermore, with the concern of construction and demolition waste remaining as a major contributor to solid waste generation, this chapter will discuss various effective waste minimisation management strategies in the construction sector.

6.2 Solid Waste Management: A Global and National View

Managing solid waste with proper disposal is a major challenge in many countries around the world. The amount of municipal solid waste generated is estimated as 2.01 billion tonnes/year in 2016, and it is projected to be 3.40 billion tonnes/year in 2050 (Kaza et al., 2018). In the previous edition of *What a Waste: A Global Review of Solid Waste Management*, it was estimated the global waste production of 1.3 billion tonnes/year in 2012 (Hoornweg & Bhada-Tata, 2012). The lower income countries suffers from increasing waste generation and improper disposal with the fact that the waste generation in these countries will be more than double in the next 20 years. From financial point of view, the solid waste management costs USD 205.4 billion to about USD 275.5 billion in 2025. Again, the cost increase will be more severe in low-income countries, compared with high-income countries. Table 6.1 shows the waste generation per capita in regions of the world. The average waste generation is very high in Organisation for Economic Co-operation and

© Springer Nature Switzerland AG 2021
X. Wang, S. Ramakrishnan, *Environmental Sustainability in Building Design and Construction*, https://doi.org/10.1007/978-3-030-76231-5_6

Table 6.1 Waste generation per capita (Hoornweg & Bhada-Tata, 2012)

Region	Waste generation per capita (kg/capita/day)		
	Lower	Upper	Average
Africa	0.09	3.0	0.65
East Asia and Pacific	0.44	4.3	0.95
Europe and Central Asia	0.29	2.1	1.1
Latin America and the Caribbean	0.11	5.5	1.1
Middle East and North Africa	0.16	5.7	1.1
OECD countries	1.10	3.7	2.2
South Asia	0.12	5.1	0.45

Development (OECD) countries, though some of the countries such as South Asia, Middle East and North America consume much higher than average.

In Australia, solid waste generation was estimated as 74 million tonnes during 2014/2015 with the per capita generation of 2.7 tonnes/year. Among the generated waste, about 27 million tonnes went to landfill and about 35 million tonnes were recycled (Pickin & Randell, 2017). The recycled waste is about 58% of the total generated waste, or equivalent to 1.47 tonnes per person. Compared to 2006/2007, the total waste generation in 2014/2015 was increased by 11% over 9 years, and waste generation per person decreased by 3% at the same time. Also, the waste that goes to landfill was dropped by 8% with the increased recycled waste by 30%. Figure 6.1 shows the waste generation trend from 2006/2007 to 2014/2015.

6.2.1 Types of Solid Wastes

Solid wastes can be generally categorised into:

- Municipal (M) waste: Wastes collected by or on behalf of municipalities. Municipal solid waste includes household and other similar waste, as well as bulky waste, yard waste, leaves, grass clippings, street sweepings and the content of litter containers. Waste from municipal sewage networks and treatment as well as municipal construction and demolition is excluded.
- Commercial and industrial (C&I) waste: Mainly the wastes that are collected from commercial buildings, government facilities, educational institutions and industrial sites
- Construction and demolition (C&D) waste: C&D waste is residential, civil and commercial waste produced by demolition and construction of buildings (though excluding most waste from owner/occupier renovations, which are usually included in the municipal waste stream).

In Australia, about 13 Mt of solid waste was collected from municipalities during 2014/2015, and about 51% of them was recovered. However, it must be stressed that this is the lowest resource recovery rate among the three main waste streams. The

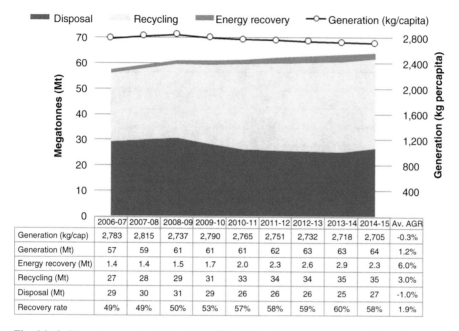

Fig. 6.1 Solid waste generation in Australia 2006–2007 to 2014–2015. (Pickin & Randell, 2017)

generation of C&I waste was about 31 Mt (20 Mt excluding fly ash), of which 57% was recovered.

The remaining 20 Mt was C&D waste, and 64% of them was recovered. In Australia, C&D recovery is well established in most states and territories, but opportunities remain for recovering material from mixed C&D waste loads, which are often taken directly to landfill. Figure 6.2 is an illustration of different waste streams in Australia.

6.2.2 Impacts of Solid Waste

Waste to landfill may have considerable impacts, not only on the aspect of land use, but also on environmental issues. For example, the biological degradation of organic wastes generates methane and contributes to greenhouse gas emissions. It may also create large impacts on local community such as uncollected solid waste that contributes to flooding, air pollution and public health impacts such as respiratory ailments, diarrhoea and dengue fever.

There are also reported health risks associated with material use, especially in building and construction. Buildings are made from a wide range of materials. Some of these materials contain chemicals that can impact health in forms such as sick house syndrome or endocrine disruption by environmental hormones. Table 6.2 lists the various harmful chemicals from C&D wastes. Categories of materials that

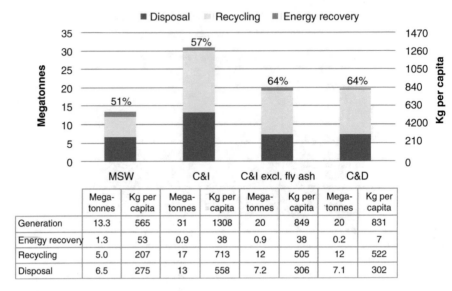

Fig. 6.2 Solid waste types and recovery rates in Australia. (Pickin & Randell, 2017)

Table 6.2 Harmful chemical substances of solid wastes

Type	Substances
Volatile organic compounds	Benzene, toluene, xylene, etc.
Organochlorines	Dioxins, trichloroethylene, etc.
Agrochemicals	Methyl bromide, chlorpyrifos, etc.
Metallic compounds	Lead and its compounds, organic tin compounds
Ozone-depleting substances	Chlorofluorocarbons, hydrochlorofluorocarbons, etc.
Others	Asbestos, etc.

have at least some probability of containing health risk chemicals are adhesives, sealants, parting agents, waterproofing agents, anticorrosion treatments, paints and undercoats.

In September 2008, leaking of methane generated from a landfill in the Cranbourne, a suburb at the fringe of Melbourne, caused the call of evacuation of local residents in a newly developed real estate area. Not only did it cause impacts on living standard of local communities, it also has potential impacts on local real estate (Jessup, 2017).

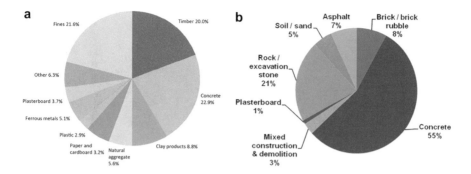

Fig. 6.3 Building waste by types in proportion of total weight in (**a**) New South Wales (**b**) VIctoria (Hyder Consulting, E.C.S.R.S., 2011).

6.2.3 Construction and Demolition (C&D) Waste

Construction and demolition (C&D) waste represents a large part of the total solid waste in most countries ranging from 20% to 60% (Akhtar & Sarmah, 2018). The main type of construction waste is soil rubble, followed by concrete-based masonry and clay-based waste such as bricks and tiles, as shown in Fig. 6.3. About 50–85% of the solid waste is concrete/masonry with the remainder being timber, metal and plastics. About one-third of the waste is reused or recycled. In Australia, the quantity of C&D waste increased from 14.9 to 19.6 Mt (from 724 to 831 kg per capita) from 2006 to 2015; however, although the generation was increased, most of the increase was recycled (Pickin & Randell, 2017). The environmental impacts associated with C&D waste disposal include land contamination, toxicity and consumption of land resources. Moreover, some types of building and construction waste have a greater environmental impact than others. For example, gypsum plasterboard disposed of in landfill produces poisonous hydrogen sulphide. This should be considered in the waste minimisation management.

6.3 Waste Treatments

The proper waste treatment strategies should follow the sequence starting from waste minimisation strategies to controlled landfill. The following list identifies the sequence of steps to be taken for proper waste treatments:

- Source reduction
- Collection
- Recycling
- Composting
- Incineration
- Landfilling/dumping

Fig. 6.4 Waste treatment hierarchy. (Hoornweg & Bhada-Tata, 2012)

The waste management should follow the hierarchical approach shown in Fig. 6.4 to identify the priority to most preferred options. The minimisation of waste generation (also known as reduce) should be approached by:

- reducing consumption of resources where possible
- reusing existing buildings and materials
- recycling resources that are left over or have reached the end of their useful life.

In addition, effective waste minimisation management strategies should involve all stakeholders (developers, architects, engineers, designers, material suppliers, waste collectors and recyclers), considering all stages in building and construction including design, construction, operation and demolition stages. Decisions on waste management should be made from the earliest stages of design.

Australian government and major companies and associations from building and construction industry have been in partnership to develop WasteWise Construction Program in a coordinated response to achieve waste reductions in 1995–1998 for Phase I and 1999–2001 for Phase II. It developed waste reduction guidelines for the construction and demolition industry as well as WasteWise Construction Handbook (Bell & McWhinney, 2000). It advocated the agenda that takes waste being resourceful, rather than wasteful. In the order of preference, the consideration in the waste management strategies should start from:

- avoid
- reduce
- reuse
- reprocess
- reclaim

- treat
- dispose

WasteWise steps for best practice include:

- Environmental policy: Confirm the organisation's commitment to environmental management and define its policy and objectives.
- Environmental/waste planning: Conduct environmental planning during design, construction and building use, including identifying legislative and regulatory requirements, identifying environmental impacts and setting waste objectives.
- Project implementation: Identify resources and responsibilities, establish project control and management by developing an administrative system and work procedures to implement the environmental objectives and establish operational procedures and controls.
- Project/program evaluation: Establish procedures to monitor, measure and record key environmental performance.
- Program revision: Modify old systems and procedures as a result of program evaluation to progressively enhance environmental performance.

The waste minimisation management strategies at every stage of building and construction are identified below:

Design stage

- Proper development/project procurement
- Design and documentation – Establish achievable levels of waste minimisation in design and documentation and tendering stages, which may affect the performance of buildings over the whole life cycle.

  - site development
  - design for greater flexibility
  - materials/component choices
  - reuse and recycling options
  - response to building codes and regulation

Construction stage

- In coordination with subcontractor and suppliers, recover waste materials by reusing and reprocessing on-site to minimise collection and/or disposal fee.
- Selling to recyclers to offset the cost of waste recovery
- Use separate waste collector for recycle waste and landfill waste at site to monitor and report diversion rates
- Having site personnel separate, monitor and report on waste diversion rate and disposing of waste to landfill when no alternative is available

Demolition stage

- Allow the choice of sequential demolition or induced collapse with consequences for reuse and recycling, i.e. a method that allows the controlled collapse of the whole or part of the building by systematically and sequentially removing the key structural members and then inducing the necessary force for demolition.

- Consider volumes of particular types of materials, large amount of material produced, distance to landfill leading to high disposal cost and requirement of scheduling, planning and safety in demolition.
- Proper waste collection: The process may include collectors, transporters, reprocessors and recyclers of waste.

6.4   Waste to Resource Management: A Circular Economy Approach

Traditionally, a linear economical approach is used when the waste management is considered. The source materials are extracted from natural resources and they are processed to make products. Once the product reaches its end of lifespan, it will be discarded to environment as a waste disposal. This linear economy approach has serious implications on the environment with the continuous increase in extraction of natural resources and generation of wastes.

The circular economy approach aimed at eliminating waste and the continual use of resources. This system encourages reuse, repair, refurbishment, remanufacturing and recycling to create a closed loop system. The World Economic Forum (Cite here: Forum) identifies the major advantages of this system as:

- Replacing the "end-of-life" concept with restoration.
- Promotes the use of renewable energy.
- eliminate the use of toxic chemicals, which impair reuse.
- Eliminate the waste through the superior design of materials, products, systems.

Figure 6.5 shows the technical nutrients and biological nutrients in the paradigm of circular economy with possible restoration methods. The concept aims to keep products, equipment and systems for longer use, thus improving the effectiveness of the resources. It also suggests that all waste should feed into another process as either a by-product or a recovered resource such that the extraction of additional resource will be limited.

6.5   Challenges with Reuse and Recycling in the Construction Sector

There are many challenges when reuse/recycle concepts are applied in building construction. The most prominent challenge being that the disassembly of buildings at the end of their life for reuse of materials and products has not been a design consideration in current practice. One of the major barriers to reuse is the difficulty of

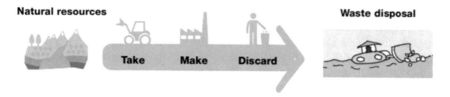

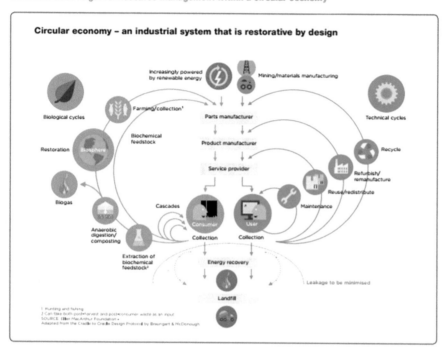

Fig. 6.5   Waste to resource management method. (Wilson et al., 2015)

separating without damaging the components that can be reused because they were not initially designed for such a purpose.

In Australia, the current building regulation does not address the question of reuse of building materials and products. The concept of reuse may be extended to in-situ building structural elements such as existing piles or perimeter walls, etc. However, the direct reuse or recycle of dissembled building products remains as a major challenge. It is also useful to predict the volume of recyclable materials at the design stage so that the estimated materials are best utilised at the end of the building life cycle.

6.6   Summary

This chapter discussed solid waste management as a growing issue in buildings and presented methods to manage solid waste. The global and Australian trend on the solid waste generation and treatment has shown that the amount of solid waste generated increases continuously, although the treatment technologies vary between countries. The classification of solid waste and their impacts on the human and built environment are also presented. The treatment of waste in the construction sector is discussed with the possible solution to apply reduce, reuse and recycle technology during the stages of design, construction and demolition of buildings. Waste to resource management using the circular economy approach has been identified as a method to reduce waste disposal to environment.

References

Akhtar, A., & Sarmah, A. K. (2018). Construction and demolition waste generation and properties of recycled aggregate concrete: A global perspective. *Journal of Cleaner Production, 186,* 262–281. https://doi.org/10.1016/j.jclepro.2018.03.085

Bell, N., & McWhinney, S. (2000). *Wastewise construction handbook: Techniques for reducing construction waste*. National Heritage Trust, Commonwealth Department of the Environment and ....

Forum, W. E. From linear to circular—Accelerating a proven concept. Retrieved from https://reports.weforum.org/toward-the-circular-economy-accelerating-the-scale-up-across-global-supply-chains/from-linear-to-circular-accelerating-a-proven-concept/

Hoornweg, D., & Bhada-Tata, P. (2012). What a waste: A global review of solid waste management.

Hyder Consulting, E.C.S.R.S., (2011). Construction and demolition waste status report. e. Department of sustainability, water, population and communitie & Q. doear management (Eds.). Melbourne, Australia.

Jessup, B. (2017). Trajectories of environmental justice: From histories to future and the Victorian environmental justice agenda. *Victoria University Law and Justice Journal, 7,* 48.

Kaza, S., Yao, L., Bhada-Tata, P., & Van Woerden, F. (2018). *What a waste 2.0: A global snapshot of solid waste management to 2050*. The World Bank.

Pickin, J., & Randell, P. (2017). *Australian national waste report 2016*. Department of the Environment and Energy.

Wilson, D. C., Rodic, L., Modak, P., Soos, R., Carpintero, A., Velis, K., Iyer, M., & Simonett, O. (2015). *Global waste management outlook*. UNEP.

Chapter 7
Sustainable Building Design

7.1 Introduction

After having a clear-cut knowledge on resource efficiency, carbon emission and environmental impacts in building and construction processes, this chapter will discuss the information that is of direct practical use for sustainable construction. It lists the key indicators that need to be considered for management to be sustainable. In practical terms, a sustainable construction is an implementation process that minimises the negative impacts on natural surroundings, with minimised consumption of natural resources and without compromising the essential needs of people through the full life cycle of buildings and infrastructures.

The implementation of sustainable construction management has to meet the objectives of key organisational, social, environmental and economic performance indicators. For example, the purpose of building construction is to improve the indoor environment to make it more suitable for human occupation. These include thermal comfort, lighting, acoustical comfort and indoor air quality. Indoor environment is therefore one of the main considerations on the aspects of building performance in relation to sustainable construction. This chapter also discusses the features of good indoor environment, assessing methods of indoor conditions and design aspects to achieve good indoor environment.

7.2 Sustainable Building Design Opportunities

The design approaches used for sustainable building construction imply various concepts including green buildings, low-energy buildings, high-performance buildings, zero-carbon development, etc. Although different concepts or definitions are introduced, they all aimed to have less negative impacts on the environment than standard buildings. The design principles often consider low resource and energy consumption

© Springer Nature Switzerland AG 2021
X. Wang, S. Ramakrishnan, *Environmental Sustainability in Building Design and Construction*, https://doi.org/10.1007/978-3-030-76231-5_7

during the construction and operation stages and generate low waste during demolition. However, the principles applied to each definition may vary with their scope. In this chapter, we will discuss three major sustainable design concepts of:

- Green building design
- Low-energy building design
- Zero-energy/zero-carbon development

7.3 Green Building Design

Green buildings take the sustainable design approach, which aims to have less negative impact on the environment than standard buildings. The design often considers construction with minimized on-site grading and natural resources by using alternative building materials and recycling construction waste rather than sending truck after truck to landfills. It also adopt interior spaces having natural lighting and outdoor views, while a superior indoor air quality is ensured by having highly efficient heating, ventilating and air-conditioning (HVAC) systems and low-volatile organic compound (VOC) materials like paint, flooring and furniture. Meanwhile, it addresses the occupants who should be healthier in the environment. For office building, it potentially boosts the productivity of workers and has lower overhead costs. Figure 7.1 demonstrates some of technical considerations in green office building.

Interior lighting: Typically, interior lighting makes up a large proportion of energy consumption of office buildings. Lighting may also generate more heats, potentially leading to more demand of air-conditioning and more energy consumption. A proper design of building orientation can effectively utilise daylight and considerably reduce artificial lighting and save energy consumption.

Orientation: Proper selection of building orientation is also useful to capture the breezes through rooftop clerestories at locations commonly subject to winds, providing cross-ventilation, and subsequently reduce energy consumption for HVAC.

Building form: A long and narrow building shape may be considered in the design to maximise natural lighting and ventilation.

Windows: Openable windows and skylights enable more natural ventilation. Windows with low-emission glazing minimize interior solar heat gain and glare.

Reflective surface: The use of reflective surfaces, mature trees to shade building walls and roofs on low-rise buildings may significantly reduce heat effect and minimize interior solar heat gain, especially in summer.

Green roof: It is particularly useful to create a green roof landscaped with drought-tolerant grasses and plants, which may lessen the heat island effect. A green roof also helps clean the air, serves as a wildlife habitat and absorbs and filters rain that would otherwise overload the storm drains and streets.

Technology: Green building technologies, including microturbines, and photovoltaic systems, reduce energy demand of grid electricity.

Water: Water-conserving irrigation systems and plumbing, waterless urinals and native and drought-tolerant landscape plants, recycled water for landscaping needs

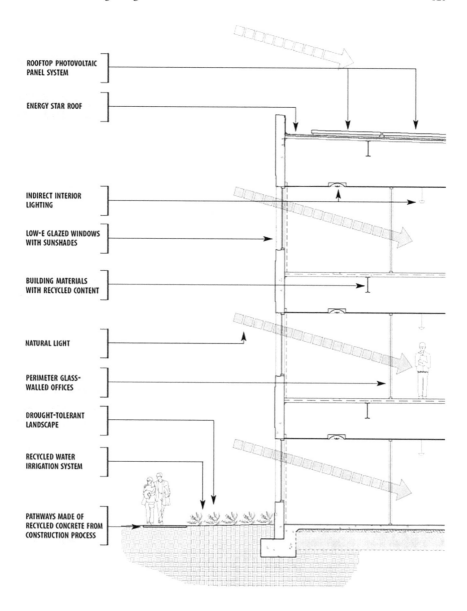

Fig. 7.1 Some design considerations in green office buildings. (Lockwood, 2006)

Recycled materials: Sustainable, nontoxic building materials are important to healthier indoor environment. These include low- and zero-VOC paints, strawboard made from wheat and linoleum flooring made from jute and linseed oil. The increase of recycled materials such as recycled carpeting and heavy steel, acoustic ceiling tiles and furniture with significant recycled content and soybean-based insulation may benefit to the reduction of resource consumption and landfill. This also includes recycling construction waste. However, green building design should balance the benefit. Increase of more windows may certainly increase the use of natural sunlight, but it also reduces the degree to which exterior walls are insulated. While more solar radiation is preferred in winter, it would be avoided in summer.

Moreover, green building design is now considered as a process, which involves owners, architects, interior/landscape designers and engineers. Integrated building design is considered to achieve overarching building design goals in terms of high-performance, low-energy, sustainable buildings. It is a collaborative process, looking into how the building and its systems can be integrated with supporting systems and how materials, systems and products connect, interact and affect one another.

7.4   Low-Energy Building Design

The conceptof low-energy building design, developed in Europe, is generally considered to be buildings that typically consume energy in the range from 30 kWh/m^2 to 20 kWh/m^2 per annum, in comparison with the average consumption of 200–250 kWh/m^2 per annum. Below this, the term 'ultra-low-energy building' is often used. When net annual energy consumption turns to zero, it is known as zero-energy building design. It should be mentioned that low-energy building design is only a design that considers energy-related indicators in sustainable building design.

Low-energy buildings mainly focus on the demand side management to reduce the annual energy consumption of the building. The general approaches used for demand reduction in low-energy buildings can be categorised into the following aspects (Kibert, 2016):

- Use building energy simulation tools for design
- Optimise the passive solar design approaches
- Maximise the thermal performance of building envelope
- Maximise daylighting and high-efficiency lighting system
- Design-efficient HVAC system
- Internal loads efficiency

The very first approach to achieve low-energy buildings is to predict the energy consumption during the design stage. The building simulations allow the user to model the building in a computer interface and assess the impact of each decision-making process on the building energy consumption. This will enable the designers to choose the appropriate construction materials, elements and the construction processes to achieve minimum energy consumption of the building. In addition, the building simulation tools allow the designers to examine the passive design concepts (which we will discuss in the next section), such as orientation and shape of the building, windows, ventilation systems, etc., to minimise the energy consumption.

Simulation tools mainly consider the thermal balance of the buildings considering external and internal thermal loads on a building. They use complex heat transfer equations to solve the heat imbalance in a building when an external or internal thermal solicitation is experienced. There are many building simulation tools available to assess the building energy consumption, particularly the operative energy consumption. Some of the widely known tools are EnergyPlus (EnergyPlus, 2012), TRNSYS (Fiksel et al., 1995), AccuRate (CSIRO) and DesignBuilder (Tronchin & Fabbri, 2008). Although the building simulation tools predict the energy consumption without much effort, the accuracy of the results should also be considered. Simulation tools require calibration, validation and verification of in-built programs that are used to solve the thermal balance algorithms. The validation can be achieved by either comparing the outputs of the simulation tool with the experimental observations or benchmarking the simulation outputs with a validated simulation tool. Most of the popular simulation tools validate their in-built programs with the experimental results on a timely manner. However, consideration of new and emerging construction materials or processes, such as dynamic thermal insulation or phase change materials, requires validation of the developed thermal model since the validation has not considered these materials.

Passive solar building design techniques are the critical technology applied to the low-energy building design. It relies on natural sources of heating, cooling, lighting and ventilation. Natural resources include sunlight, wind, vegetation, etc. It also defines the energy character of the building before considering active systems to meet thermal comfort in buildings. The maximum benefits that can be obtained by passive design techniques are mainly governed by the building location/site, surroundings and the building design.

Passive solar building includes six distinct design elements as listed below (Kibert, 2016):

- Building orientation and aspect ratio
- Building envelope design
- Daylighting strategies
- Ventilation strategies
- Thermal mass and insulation
- Internal load reductions

7.4.1 Building Orientation and Aspect Ratio

The orientation of the building plays a major role in the energy performance of buildings. Orientation of a building means the positioning in relation to the seasonal variations in the sun's path as well as prevailing wind patterns. The location of the building or local climate conditions significantly influence the building orientation and its aspect ratio. Figure 7.2 shows two houses located in different climate regions of temperate climate zone and tropical climate zone. It can be seen that the house located in temperate climate regions requires more openings (i.e. windows) in the north direction (it is only applicable to southern hemisphere like Australia; in northern hemisphere, it is south direction) to receive direct solar radiation for heating in winter. Furthermore, the openings in west direction should be minimised to prevent heat gain during summer evening.

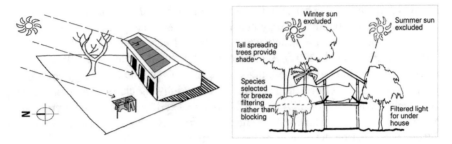

Fig. 7.2 House in (**a**) temperate climate zone (**b**) tropical climate zone (Reproduced from Peter Lyons et al., 2013)

On the other hand, in tropical climate regions (countries close to the equator), the path of the sun changes gradually throughout the year between summer and winter. Therefore, buildings should be oriented such that the majority of the walls and windows can be easily shaded from the direct sun. Depending on the building use, it may be desirable to admit some northern sun (for countries in southern hemisphere) in the winter months, which can be done by planning the width of eaves and awnings. Growing tall spreading tropical trees can be advantageous to provide shade to the building.

The site conditions also have an impact on the passive design. If the site has right characteristics, good passive solar performance can be achieved with a minimal cost. For example, if the house is surrounded by tall buildings or shade-giving plants, it is necessary to choose optimum daylighting and natural ventilation strategies as discussed below.

7.4.2 Building Envelope Design

The building envelope system contributes to the energy transmission through the building skin as conduction, convection and radiant heat transfer processes. The major envelope surfaces are walls, windows and roof. Efficient design and construction of building envelope systems could substantially reduce the energy gain into the building during summer season and energy losses during winter season. The features of an efficient building envelope systems are:

- Capability to control solar heat gain into the building
- Minimise direct heat transmission into the building
- Minimise the infiltration or leakage by having a tight and thermally resistant envelope

Building Envelope Design: Wall System

Walls cover a large amount of building surface area and are responsible for direct heat transmission into the building. Building walls should be designed to have high thermal resistance and thermal mass, and their orientation significantly influences effectiveness in reducing the heat transmission into the building. The difference between thermal mass and thermal resistance is that the thermal mass has the ability to absorb and store heat energy, whereas thermal resistance delays the heat transfer through the element. Studies have reported that placing insulation materials to the exterior of the building walls and thermal mass closer to the interior would provide an effective solution for reducing the heat transmission into the building. However, it must be noted that the design of building walls also depends on local climatic conditions, and thermal simulations should be carried out to identify the most efficient wall design. For instance, the countries experiencing large diurnal temperature fluctuations can take advantages of high thermal mass as they reduce the indoor temperature fluctuations by storing excess thermal energy and discharge when required. In contrast, tropical countries with low diurnal temperature variation require high thermal insulation to delay the heat penetration into the buildings. The exterior wall surface properties also play a major role in resisting or reflecting the solar thermal radiation into the building. Building walls with high roughness can reflect more heat to the surrounding as long-wave radiation and reduce the heat penetration into the building. Light-coloured surfaces also absorb less amounts of the solar radiation compared with dark-coloured surfaces.

Building Envelope Design: Windows

Windows are meant to allow light into the room spaces, while openable windows exchange the indoor air with outdoors. Due to the requirements of light penetration through the windows, they are designed to be transparent. The transparent nature would allow solar irradiance into the rooms and easily heats up the rooms. On the other hand, window glasses are generally thinner compared with other building elements, and therefore, thermal resistance is poor. High-performance windows should incorporate the permission of light into the structure while controlling solar heat gain and conduction energy through the assembly.

The performance of windows is measured by thermal conductance (U-value), solar heat gain coefficient (SHGC) and visible transmittance. SHGC is simply a measurement of the amount of solar irradiance penetrated through the glass, as a fraction of total solar irradiance that falls on the window surface. The lower the SHGC, the less solar heat penetrates through the glazing from the exterior to the interior. It is recommended to have high SHGC glazing for north-facing windows (for the houses located in southern hemisphere) to receive winter solar radiation into the rooms. On the other hand, west-facing windows should have low SHGC to prevent the solar energy penetration during afternoons and hot summer days. Table 7.1: shows the window glazing parameters for some of the commonly used windows in Australia. Low-emissivity glasses or applying reflective coatings to normal glasses can control the SHGC of windows; however, the effectiveness of those coatings largely depends on the thickness and reflectivity of the layers.

Table 7.1 Characteristics of window types

Window type	Visible light transmittance	U value	Solar heat gain coefficient (SHGC)
Single-glazed aluminium window with 3 mm glazing	0.80	6.9	0.77
Single-glazed timber / uPVC window with 3 mm glazing	0.72	5.5	0.69
Double-glazed aluminium window with 3 mm glazing – 6 mm airgap – 3 mm glazing	0.72	4.2	0.69
Double-glazed timber/uPVC window with 3 mm glazing-6 mm airgap – 3 mm glazing	0.65	3.0	0.61

Reproduced from Peter Lyons et al. (2013)

Building Envelope Design: Roof

Roof is a major area for heat transmission due to its generally large size and direct exposure to the sun. Roof surface of the open area buildings can reach up to 83 °C in the summer days, which not only heats up the indoor of the buildings, but substantially affects the neighbourhood as low-wave radiations. The thermal performance of building roof designs can be achieved by following physical principles as mentioned in Al-Obaidi et al. (2014).

- Reflective cooling techniques by slowing down the heat transfer through roofs
- Radiative heat technique by removing unwanted heating from the building

Reflective roof designs aim at reducing the heat gains on the building roofs by carefully choosing the roof surface properties to act as a reflector of invisible electromagnetic radiations (short-wave and long-wave radiations) and as a good emitter of heat. Two key features of the reflective roofs are solar reflectance (also known as albedo effect) and thermal emittance as shown in Fig. 7.3. Albedo is a measurement used for measuring the reflectivity of the solar radiation, and a high albedo can assist in reducing the thermal loads onto the neighbourhood. Light-colour roofs have high reflectivity or albedo, and they can significantly reduce the thermal loads onto the neighbourhood. A study on the effect of roof surface reflectivity on the building energy consumptions reported that the light-coloured, reflective roof surfaces use 40% less energy than dark roofs. Another study conducted experiments on three identical buildings using three different coating materials and reported that the increase in surface reflectance from 32% to 61% can reduce the annual energy consumption by 116 kWh (Shen, 2011). Different types of reflective roofs according to the slope of the roof summarised by Urban and Roth (2010) are given in Table 7.2:

The radiative cooling technique utilises the emissivity characteristics of the roof surfaces to emit the energy in the form of electromagnetic radiation. Generally,

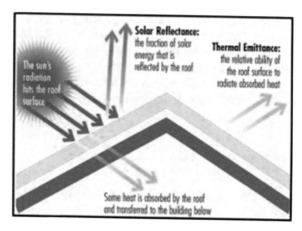

Fig. 7.3 Solar reflectance of roofing (Cushman, 2020)

Table 7.2 Different types of reflective roofs (Urban & Roth, 2010)

Roof type	Roofing material properties
Flat roofs	
Cool roof coatings	Roof surfaces are painted with white or other special pigments that can reflect more solar radiation. The coatings are made thicker to avoid ultraviolet deterioration and chemical damages
Low sloped roofs	
Single-ply membranes	A membrane of highly reflective prefabricated sheets is installed above the rooftop using mechanical fasteners or chemical adhesives
Built-up roofs	A sheet layer, composed of a base sheet, fabric reinforcement layers and a dark preservative surface layer, is attached on the roof surface
Modified bitumen sheet membranes	One or more layers of reinforced plastic or rubber and covered with sooth finish or mineral granules
Spray polyurethane foam roofs	A strong chemical is sprayed on the roof surface
Steep-sloped roofs	
Shingled roofs	Overlapping panels are made of specific materials; fibreglass asphalt shingles are most commonly used in residential buildings
Tile roofs	Tiles are made of concrete/clay/slat and depending on their properties; the surface colours also vary and hence the reflective cooling properties
Low- and steep-sloped roofs	
Metal roofs	Metal roofing materials are obtained through granular-coated surfaces and are having natural metallic finishes or oven-baked paint finishes

radiative heat flux occurs when the two surfaces are at different temperatures and face each other. The radiative heat transfer occurs until an equilibrium state is reached. For buildings, the radiative cooling is based on the temperature difference between sky conditions and the building surfaces, whereas cloud cover, air humidity and pollution could reduce the cooling performances. The radiative cooling technique can be beneficial in both day and night. During the daytime, roof surfaces can absorb the excess heat from the rooms below and losing the heat through long-wave radiation during night where the ambient temperature is lower than in roof surfaces. The strategy of choosing coloured roofs as well as the lightweight structures influences the long-wave radiation emission during the night-time, in addition to the reflection of solar irradiance during daytime.

7.4.3 Daylighting Strategies

Natural light or daylighting not only supplies lights to indoor for free; they have also shown to provide physical and psychological benefits to the occupants. Almost every building can benefit from daylighting to a certain extent, if proper strategies are applied in installing the daylighting devices. Some of the key considerations in daylighting are accessibility of light through windows, the selection of glazing type, optimum daylighting requirement and its position as well as automated daylight-activated controls. Windows are the primary elements to receive daylighting, and it can be achieved by positioning windows in a perimeter wall. Roof space, core, etc. depending on the building surroundings and shading. For instance, a building in high-density urban site may not receive sufficient light to fully utilise the daylighting.

When the daylighting cannot be achieved by simple windows, skylights can be used. Skylights can admit as much as three times of amounts of lights than vertical windows of the same size and distribute the light evenly into the indoor space. They are a good alternative when the lighting on the building is limited or when the building is restricted by the size of windows due to privacy concerns or architectural preferences. Figure 7.2 shows few different types of skylights available in Australia.

7.4.4 Ventilation Strategies

Ventilation strategies provide cooling of the buildings through heat loss by exchanging the warm indoor air with cooler external air. In addition to the heat loss, the ventilation improves occupant thermal comfort by increasing the evaporation from the body due to the continuous movement of indoor air. Ventilation in buildings can be achieved by active or passive methods. Passive ventilation methods include cross natural ventilation by strategically positioning windows, thermal chimney effect, Venturi effect, wind catchers, etc. Active ventilation methods use the energy to mechanically force the air exchange between indoor and outdoor. Fans and dampers use electrical energy to provide mechanical ventilation in buildings.

Narrow or open plan building layouts provide better cross-natural ventilation effects on building spaces as shown in Fig. 7.4. They work well when there is an air-pressure differential caused by wind or breezes. However, cross-natural ventilation may not be effective in buildings with following situations:

Fig. 7.4 Cross-ventilation in buildings. (Source: hubpages.com)

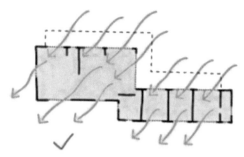

A well designed home with single room

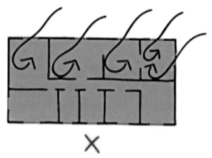

A poorly designed home that will create hot stagnant areas.

- Buildings located in high noise, security risk and poor outdoor air quality areas, where the windows need to be closed at most
- Consecutive days of high outdoor ambient temperature
- Long air movement paths due to deep spaces

7.4.5 Thermal Mass and Insulation

In passive solar building designs, thermal masses of the building components such as walls and roof act as the thermal energy storage materials to store excess solar thermal energy and discharges when required. This helps to reduce the indoor temperature fluctuations, reducing the peak indoor temperature, and also helps to shift the peak energy demand for mechanically cooled buildings. The reduction and shift in the peak energy demand are particularly advantageous to reduce the high demand of electricity grid during the peak times, allowing to choose a smaller capacity air-conditioning systems, and operate the air-conditioning system at an efficient mode because the peak load occurs later in the evening when the outdoor air temperature dropped from midday. Figure 7.5 shows the indoor temperature fluctuations of two cases: one with low thermal mass and the other with high thermal mass. It can be seen that the high thermal mass significantly reduces the diurnal temperature

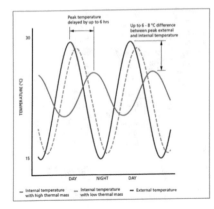

Time lag figures for various materials	
Material thickness (mm)	Time lag (hours)
Double brick (220)	6.2
Concrete (250)	6.9
Autoclaved aerated concrete (200)	7.0
Mud brick/adobe (250)	9.2
Rammed earth (250)	10.3
Compressed earth blocks (250)	10.5

Fig. 7.5 Effect of thermal mass on indoor temperature. (Reardon et al., 2013)

fluctuations and has high peak load shifting. High thermal mass is required for temperate climates where a significant diurnal temperature variation is experienced. During summer time, the thermal mass stores the heat energy during daytime and is discharged during cooler night-times. Also, the cold storage during night can alleviate the hot indoor environment during daytime. In tropical climates, however, the use of lightweight construction with low thermal mass is preferable. This is because the tropical climates experience low diurnal temperature fluctuations with warm nights. Lightweight construction materials such as timber do not have high thermal mass and respond to the evening/night cooling breezes quickly.

In addition to the thermal masses, thermal resistance (insulation) of building components is important to reduce the transmittance of heat energy from outdoor to indoor and vice versa. The thermal resistance of buildings is controlled by the insulation material properties and thickness. It is worth mentioning here that generally insulation materials are not good at thermal energy storage; nevertheless, they also play a significant role in improving the indoor thermal comfort in buildings.

The common insulation material applications in buildings are walls, ceiling, roof and underfloor insulations. There are generally two types of insulation products available: bulk insulation and reflective insulation. While bulk insulation is good for both keeping warmer air in during the winter and preventing hot air in the building during summer, reflective insulations are highly effective for hot summer days. The insulation capacity of building materials is measured by its 'R-value'.

A high-performance building component is characterised by its thermal storage medium and insulation layers as well as their distribution or relative locations. Accordingly, it is important to understand the effect of thermal mass and insulation separately as well as a single system, where either of the materials affects the performance of other. A case study on different construction types of exterior wall system was conducted using building simulations to understand the best combination or distribution of thermal mass and insulation in building walls (Gregory et al., 2008). Four different wall construction types as shown in Fig. 7.6 are considered for the analysis. The temperature profiles of two modular houses (Module A: 6×4 m^2

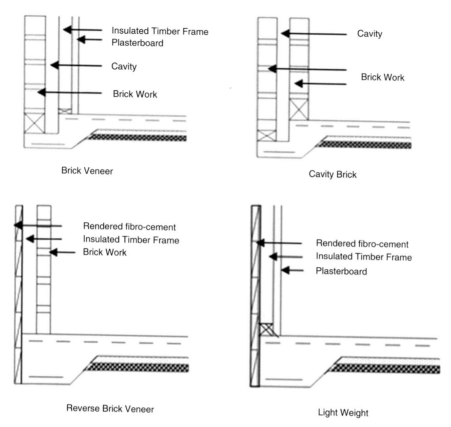

Brick Veneer

Cavity Brick

Reverse Brick Veneer

Light Weight

Fig. 7.6 Effect of construction systems on the performance of buildings. (Gregory et al., 2008)

room; Module B: 6 × 4 m² room with 5 × 2 m² window) are reported in Fig. 7.7. Among the different construction types, reverse brick veneer construction has shown lowest diurnal indoor temperature fluctuations followed by brick veneer, cavity brick and lightweight construction systems. Thus, the reverse brick veneer construction provides the least energy consumption in buildings for the analysed climate zone and indicates that the thermal mass performs well within a protective insulation envelope.

In addition to the traditional construction systems, advanced construction systems to increase the thermal mass and insulation capacity of the building envelope were explored. For instance, a Trombe wall shown in Fig. 7.8 (a) contains a high thermal mass wall protected by a glazing system. The air in the space between thermal storage wall and glazing is heated up by the winter solar irradiance. As the air is heated, it passes through the top vent and circulates the room by supplying the cold air through bottom vent to the Trombe wall space. This would create a cycle of warm air flow. While the warm air circulation heats up the room through convection, the night thermal comfort is achieved by the discharge of heat from the thermal

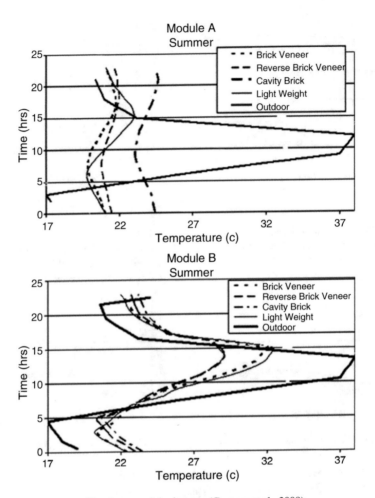

Fig. 7.7 Temperature profile of two modular houses. (Gregory et al., 2008)

storage wall. This will provide all-day warmth and comfort in the room. Trombe walls are generally placed in the north-facing direction of the house (in Australia) to harness maximum solar radiation during winter as the sun travels at a low altitude (Fig. 7.8b).

Solar chimney is another type of passive heating and cooling system that regulates the indoor air temperature and provides ventilation (Fig. 7.9). The process of heating a room using solar chimney is similar to the Trombe wall, where the air inside the column is heated up and circulated to the room via the top vent. This will pull back the cold air into the chimney, heating the air again. Cooling a space using solar chimney is achieved by the cross-ventilation circulating the outdoor air through the vent in the opposite end of the building to the top vent in the chimney. This air circulation creates a sort of 'draft' effect that provides cool, fresh air into the building (Bansal et al., 1993).

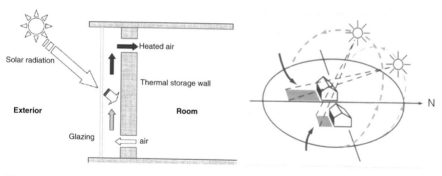

Fig. 7.8 (**a**) typical Trombe wall construction (Quesada et al., 2012) (**b**) the sun's path during summer and winter in Australia. (Peter Lyons et al., 2013)

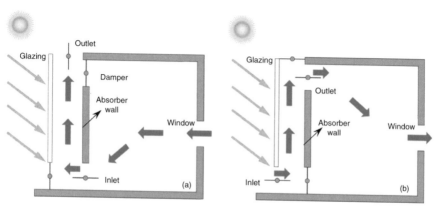

Fig. 7.9 Typical wall solar chimney under: (**a**) cooling mode and (**b**) heating mode. (Shi, 2018)

7.4.6   Internal Load Reduction

Apart from external heat gain opportunities, the internal heat loads are also another form of thermal loads on the building, and a significant effort is required to reduce heat gains from internal loads. Efficient daylight strategies can partially reduce the lighting energy consumptions as well as the heat gains from lighting equipment. However, there any many other sources that have comparably high heat gains than lighting in buildings. People are a major contributor to internal heat gain as the human body discharges heat to the environment based on their work. Table 7.3 shows the heat dissipation from human body for their work nature. Reducing the number of people or their work type is not a viable option. The other forms of internal heat gains are the machineries including computers, printers, scanners and other peripherals in office buildings. Factories or manufacturing industrial buildings have many machineries that dissipate heat due to the inefficiency or poor maintenance.

The internal gains due to the appliances working throughout the day such as desktop computers, printers, scanners, etc. not only consume a substantial amount

Table 7.3 Metabolism rate for various activities (Ashrae, 2009)

Activity	Activity level W/m²	met[a]
Resting		
Sleeping	40	0.7
Reclining	45	0.8
Seated, quiet	60	1
Standing, relaxed	70	1.2
Walking (on level surface)		
3.2 km/h (0.9 m/s)	115	2
4.3 km/h (1.2 m/s)	150	2.6
6.4 km/h (1.8 m/s)	220	3.8
Office activities		
Reading, seated	55	1
Writing	60	1
Typing	65	1.1
Filing, seated	70	1.2
Filing, standing	80	1.4
Walking about	100	1.7
Lifting/packing	120	2.1
Miscellaneous occupational activities		
Cooking	95–115	1.6–2.0
Housecleaning	115–200	2.0–3.4
Seated, heavy limb movement	130	2.2
Machine work	105	1.8
Sawing (table saw)	115–140	2.0–2.4
Light (electrical industry)	235	4
Handling 50-kg bags	235	4
Pick and shovel work	235–280	4.0–4.8
Miscellaneous leisure activities		
Dancing, social	140–255	2.4–4.4
Calisthenics/exercise	175–235	3.0–4.0
Tennis, singles	210–270	3.6–4.0
Basketball, competitive	290–440	5.0–7.6
Wrestling, competitive	410–505	7.0–8.7

[a]Note: one met = 58.1 W/m²

of energy but also cause heat emission. These devices require continuous running as they have long start-up time, and waiting for start-up every time makes the staffs inconvenience. One alternative is to replace the desktop computers with laptops that can change the mode to standby or sleep when not in use. They also have a very short reboot time and consume very low energy than desktop computers. Modern copying machineries also have standby mode to reduce their operation when not in use. For equipment that has limited usage time, such as microwave ovens, the energy consumed when they are not in use can be more than the energy consumption during the operations since they are only used for very short durations; however, they are

Fig. 7.10   Smart energy management system in AusZEH (Brochure)

running 24 h, 7 days per week. Timer-based power points are installed in new buildings to power off the equipment when they are not in use.

Smart energy management system can substantially control the equipment and lighting for their unwanted usage as well as the occupant behaviour–related operations. With the advancement of wireless personal area network and wireless sensor networks, control over lightings and appliances becomes far more effective. Occupants can control their lightings and HVAC systems using the mobile phones when they are on the bed or outside home and forgot to turn off the devices. The following are some advantages of smart energy management systems (Fig. 7.10):

- It tracks all energy and water use and supply and displays on a touchscreen (can be accessed remotely).
- Lighting, air-conditioning and other appliances are also able to be controlled through the touchscreen.
- Automatically switch devices on and off according to a variety of operating schedules.
- Indicating energy-saving options: reducing standby power, effective usage of heating and cooling and lighting systems in the house

7.5   Zero-Energy/Zero-Carbon Design

The concept of zero-energy design or zero-carbon design has gained a significant attention in the last decade, and it is no longer perceived as a concept with more and more demonstration projects that have been shown in the recent years. Figure 7.11 shows the number of ZEBs around the world with their corresponding climate types. More than 280 demonstration projects have been implemented around the world with their ongoing monitoring of the performances. Before moving into detail, it is necessary to understand the meaning of a ZEB.

Zero-energy building or zero-carbon building is a complex concept that considers multiple parameters/metrics to assess the zero design, and based on the metric used, the definitions also vary. A very simple diagram illustrating on how the zero-carbon design is achieved is shown in Fig. 7.12. Here carbon can be replaced by

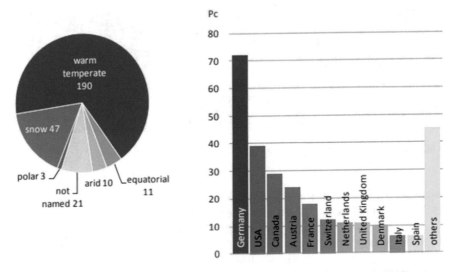

Fig. 7.11 Climate type and distribution of ZEB in various countries. (Musall et al., 2010)

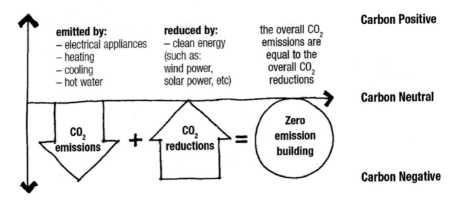

Fig. 7.12 Zero design concept in buildings

energy to attain the zero-energy design. In any case, the zero-energy/zero-carbon design is a way of balancing the energy consumption of the building with same or more amount of green energy. Although this seems a straightforward assessment, there are many parameters that vary and the corresponding definitions of zero design also vary. The way of assessing zero design can be different by the consideration of:

1. Metric of the balance: The metric used to assess the ZEB can be end-use energy, CO_2 emission, CO_2-equivalent emission, energy cost, exergy, etc.
2. The type of energy used for the analysis: The demand type that requires balancing using renewable energy can be thermal energy for HVAC operation, operative energy or life cycle energy including embodied energy.

3. The balancing period: The period of time over the energy consumption is balanced using the renewable energy. The period can be at all times, annually or
   lifespan of the building.
4. The type of energy balance: This criterion is relevant to the grid-connected ZEBs
   as to choose between two possibilities: (1) the energy use of the building is balanced by the renewable energy generation or (2) the energy delivered to the
   building is balanced by the energy feed into the grid. The main difference
   between two choices is that the first balance mechanism should be done during
   the design phase of the building and second balance mechanism in the operational phase.
5. The renewable energy option: The green energy for balancing the building
   energy consumption can be green energy purchased from grid, off-site renewable energy generation or on-site renewable energy generation. Figure 7.13
   shows the different types of renewable energy systems considered for zero design.
6. Connection to the energy infrastructure: Depending on whether the building is
   connected to the grid or not, the ZEB varies as on-grid ZEB or off-grid ZEB. Both
   on-grid and off-grid ZEBs can perform absolute ZEBs or net ZEBs (the definition of net ZEB and absolute ZEB are discussed below).

• Table 7.4 summarises the different criteria for defining a ZEB using the above-
  mentioned factors. Based on the chosen criteria, the definition of ZEB also varies. The initial attempts towards zero development were focused on thermal

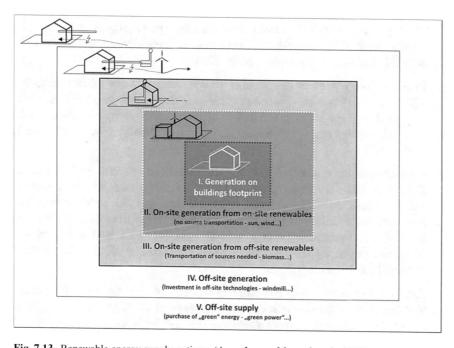

Fig. 7.13 Renewable energy supply options. (Anna Joanna Marszal et al., 2010)

Table 7.4 Metrics of balance in ZEB

Metric of balance	Demand type	Period of balance	Type of energy use	Connection to grid
Carbon emission	Thermal energy	At all times	Green energy	Off-grid
Energy	Operative energy	Annually	Off-site renewable energy	On-grid
Energy cost	Life cycle energy	Lifespan	On-site renewable energy	

energy demands during the annual operation of buildings. More precisely, the annual thermal energy demands for space heating, and hot water supply is required to be offset by the renewable energy generation during the year. Esbensen and Korsgaard defined the zero-energy house as:

- *Zero Energy House is dimensioned to be self-sufficient in space heating and hot-water supply during normal climatic conditions.* (Esbensen & Korsgaard, 1977)

- Later, the demand type was extended to the annual electricity consumption of the house including, space heating/cooling, hot water supply and appliances' energy consumption. Gilijamse defined the ZEB as follows:

- *A zero-energy house is defined as a house where no fossil fuels are consumed, and annual electricity consumption equals annual electricity production.* (Gilijamse & Boonstra, 1995)

- More recently, the ZEB is defined to consider life cycle energy consumption of a building to be offset by the renewable energy generated during the lifetime of the building. Hernandez Kenny defined the ZEB as:

- *Primary energy used in the building in operation plus the energy embodied within its constituent materials and systems, including energy generating ones, over the life of the building is equal to or less than the energy produced by its renewable energy systems within the building over their lifetime.* (Hernandez & Kenny, 2010)

- Depending on the time of balance, ZEBs can be defined as absolute ZEBs or net ZEBs. Net ZEBs are engaged with one or more energy infrastructures, such as electricity grid, district heating and cooling system, gas pipe network etc. They also have the capability of both supplying excess energy from the infrastructure to the grid and purchasing energy from the grid, thus avoiding the requirement of on-site electricity storage (Anna Joanna Marszal et al., 2011). Laustsen defined the net ZEB as follows:

- *Zero Net Energy Buildings are building that over a year are neutral, meaning that they deliver as much energy to the supply grids as they use from the grids. Seen in these terms they do not need any fossil fuel for heating, cooling, lighting or other energy uses although they sometimes draw energy from the grid.* (Laustsen, 2008; A. J. Marszal et al., 2011)

- The stand-alone ZEB or absolute ZEB is another form of achieving zero energy in a building without consuming the electricity from grid. They entirely rely on the generation and storage of energy from the renewable energy systems. Laustsen defines the stand-alone ZEB as:

- *Zero Stand Alone Buildings are buildings that do not require connection to the grid or only as a backup. Standalone buildings can autonomously supply themselves with energy, as they have the capacity to store energy for night-time or wintertime use.* Laustsen (2008)

- Stand-alone ZEBs are mainly considered for rural/remote areas where the connection to energy infrastructure is nearly impossible or incurs huge cost. They also require different forms of renewable energy systems as one form does not meet the required energy demand due to climate uncertainty. For instance, if the building is relied on solar photovoltaics (PVs) as the primary energy source and there is no enough sunlight for many consecutive days, the other forms of renewable energy systems are required. The limitations of stand-alone ZEBs are the requirements of large storage capacity, backup generators and energy losses during the storage and conversion back to the usage as well as the oversized renewable energy–producing system to meet the peak energy demands.

Figure 7.14 shows the main aspects of a ZEB as the demand side management to reduce the energy consumption of the building followed by the renewable energy generation to meet the minimised energy consumption of the building. For example, choosing the airtight super-insulated envelope, solar shading, ventilated sunspace and green roof are means of reducing heating/cooling energy demand of the

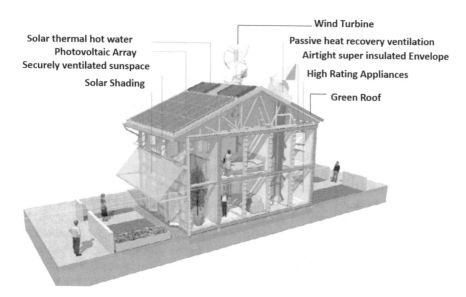

Fig. 7.14   ZEB opportunities in detached houses. (Source: RuralZED)

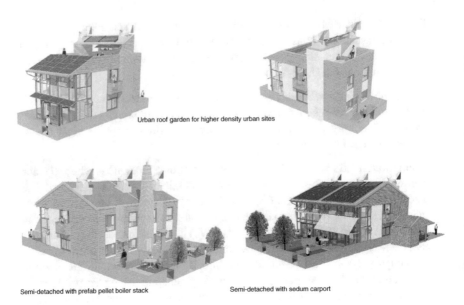

Urban roof garden for higher density urban sites

Semi-detached with prefab pellet boiler stack   Semi-detached with sedum carport

Fig. 7.15 ZEB opportunities in urban houses. (Source: RuralZED)

building. High-rating appliances reduce the internal heat gains and energy consumption due to appliances. Meanwhile, solar thermal hot water, photovoltaic array and wind turbine are to generate the renewable energy supply to the building. Achieving the ZEB in an urban residential or commercial building is quite challenging as the site area and the technologies are limited. Figure 7.15 shows few opportunities of urban ZEB designs. Due to the very limited land size, the rooftop gardens can be advantageous to save space as well as to improve the energy efficiency of the building. Heating and cooling of buildings contribute to more than half of the operative energy consumption, and having a pallet boiler stack using wood as fuel can substantially reduce the heating energy demand of the house during winter. Sedum carports are another example of incorporating green roof in the building site.

Another way of implementing ZEB is by refurbishing the existing building to meet the requirements of zero-energy design. This is particularly important as the existing buildings contribute approximately 98% of the Australian housing stock and, thus, have a great potential for improvements. Also, refurbishing the existing buildings can reduce the resource consumption for new builds and waste generated during the demolition of the existing buildings. However, the refurbishment to meet the zero-energy design is much more challenging task than the construction of a new building due to a number of obstacles that significantly can narrow down the possible technical solutions especially in the dense city area or for multi-storey buildings. Moreover, detached buildings also have challenges due to their poor design to have limited passive design opportunities. For instance, the existing houses may have poor orientation of the building that cannot be altered during the refurbishment. Furthermore, the building context and its location do not allow to

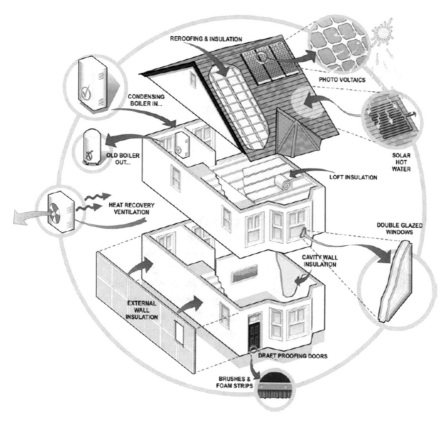

Fig. 7.16 Application of ZEB concept in existing houses (ESD, 2004)

design the building only with solar PV system–based renewable energy source as the capacity may exceed the possible installation coverage area. Therefore, additional renewable energy systems such as wind turbines and green energy purchase from grid should be considered. Another important aspect of ZED in existing houses is the balancing period. Often the life cycle of the building cannot be considered for ZED in existing buildings because very limited data are available on the embodied energy and embodied carbon of materials used in the old building. Also, the construction process may have been conducted in an inefficient way, and the data are insufficient. It is possible to consider the balancing period of annual zero-energy design. This can be achieved by balancing the annual energy consumption of the building using the renewable energy generated during the same period. Therefore, it can be concluded that ZEB concept should be different for existing and newly constructed buildings in order to make it feasible for both cases. Figure 7.16 shows the opportunities of ZED in existing buildings with the consideration of various demand side strategies and renewable energy generation technologies. Some of the demand side management strategies are:

- Increasing external wall insulation and cavity wall insulations
- Adding loft insulation to the ceiling and high reflective insulation for roofing
- Double-glazed windows for maximum protection of heat loss through glazing
- Replacing old inefficient boilers with condensing boilers
- Heat recovery ventilation systems to prevent heat loss while achieving the adequate ventilation for the building
- Draft proofing doors and foam strips installation for gaps to prevent the heat loss due to infiltration/leakage

7.6   Summary

This chapter provided the implication methods to achieve sustainable building design with the examples to show various strategies supporting these methods. The topics covered under the sustainable building designs include green building design, low-energy building design and zero-energy building design. The strategies discussed under these topics can be very useful for engineers and students to understand these methods and to develop new technologies to cater these strategies. While these topics mutually share many strategies, they also have the exclusive criteria to meet the requirements. Zero-energy buildings are a special class of buildings with the expectation to achieve net zero energy consumption during the operation. The definitions of zero-energy buildings vary as per the considered parameters such as metric of balance, demand type, balancing period and type of energy use. Finally, the opportunities to achieve zero-energy design in existing buildings were discussed with the challenges.

References

Al-Obaidi, K. M., Ismail, M., & Abdul Rahman, A. M. (2014). Passive cooling techniques through reflective and radiative roofs in tropical houses in Southeast Asia: A literature review. *Frontiers of Architectural Research, 3*(3), 283–297. https://doi.org/10.1016/j.foar.2014.06.002

Ashrae, A. H. F. (2009). American society of heating, refrigerating and air-conditioning engineers. Inc., Atlanta

Bansal, N., Mathur, R., & Bhandari, M. (1993). Solar chimney for enhanced stack ventilation. *Building and Environment, 28*(3), 373–377.

Brochure. *Australian zero emission demonstration house* [Press release]. Retrieved from www.auszeh.org.au

CSIRO. *AccuRate: helping designers deliver energy efficient homes.* Retrieved from https://www.csiro.au/science/AccuRate.html

EnergyPlus. (2012). *Engineering reference handbook.*

Esbensen, T. V., & Korsgaard, V. (1977). Dimensioning of the solar heating system in the zero energy house in Denmark. *Solar Energy, 19*(2), 195–199.

ESD. (2004). Low Carbon Homes: towards zero carbon refurbishment. Feasibility study for the EnergySaving Trust Innovation Programme.

Fiksel, A., Thornton, J., Klein, S., & Beckman, W. (1995). Developments to the TRNSYS simulation program.

Gilijamse, W., & Boonstra, M. (1995). Energy efficiency in new houses. Heat demand reduction versus cogeneration? *Energy and Buildings, 23*(1), 49–62.

Gregory, K., Moghtaderi, B., Sugo, H., & Page, A. (2008). Effect of thermal mass on the thermal performance of various Australian residential constructions systems. *Energy and Buildings, 40*(4), 459–465. https://doi.org/10.1016/j.enbuild.2007.04.001

Hernandez, P., & Kenny, P. (2010). From net energy to zero energy buildings: Defining life cycle zero energy buildings (LC-ZEB). *Energy and Buildings, 42*(6), 815–821.

Kibert, C. J. (2016). *Sustainable construction: Green building design and delivery.* Wiley.

Laustsen, J. (2008). Energy efficiency requirements in building codes, energy efficiency policies for new buildings. IEA Information Paper. *Support of the G8 Plan of Action.*

Lockwood, C. (2006). Building the green way. *Harvard Business Review, 84*(6), 129–137.

CUSHMAN, T. (2020). MANAGING SOLAR GAIN [Online]. Builder Online. Available: https://www.builderonline.com/building/building-science/managing-solar-gain_o [Accessed on 13/02/2021].

Marszal, A. J., Bourrelle, J. S., Musall, E., Heiselberg, P., Gustavsen, A., & Voss, K. (2010). Net zero energy buildings-calculation methodologies versus national building codes. *The Proceedings of EuroSun.*

Marszal, A. J., Heiselberg, P., Bourrelle, J. S., Musall, E., Voss, K., Sartori, I., & Napolitano, A. (2011). Zero energy building – A review of definitions and calculation methodologies. *Energy and Buildings, 43*(4), 971–979. https://doi.org/10.1016/j.enbuild.2010.12.022

Musall, E., Weiss, T., Lenoir, A., Voss, K., Garde, F., & Donn, M. (2010). *Net Zero energy solar buildings: An overview and analysis on worldwide building projects.* Paper presented at the EuroSun conference.

Peter Lyons, B., Chris Reardon, C., & Tracey Gramlick, R. (2013). *Passive design, glazing—Your home: Australia's guide to environmentally sustainable homes.* Commonwealth of Australia (Department of Industry).

Quesada, G., Rousse, D., Dutil, Y., Badache, M., & Hallé, S. (2012). A comprehensive review of solar facades. Opaque solar facades. *Renewable and Sustainable Energy Reviews, 16*(5), 2820–2832.

Reardon, C., Downton, P., & McGee, C. (2013). *Materials, construction systems—Your home Australia's guide to environmentally sustainable homes.* Commonwealth of Australia (Department of Industry).

Shen, H., Tan, H., & Tzempelikos, A. (2011). The effect of reflective coatings on building surface temperatures, indoor environment and energy consumption—An experimental study. *Energy and Buildings, 43*(2), 573–580. https://doi.org/10.1016/j.enbuild.2010.10.024

Shi, L. (2018). Theoretical models for wall solar chimney under cooling and heating modes considering room configuration. *Energy, 165*, 925–938. https://doi.org/10.1016/j.energy.2018.10.037

Tronchin, L., & Fabbri, K. (2008). Energy performance building evaluation in Mediterranean countries: Comparison between software simulations and operating rating simulation. *Energy and Buildings, 40*(7), 1176–1187.

Urban, B., & Roth, K. (2010). *Guidelines for selecting cool roofs.* US Department of Energy.

Chapter 8
Resilience and Adaptation in Buildings

8.1 Introduction

Despite the fact that energy efficiency in buildings is an important feature of sustainable development, its operative performance to the changing climate should also be ensured. Buildings are designed to operate for longer periods, preferably exceeding 40–50 years, and the initial design of buildings should enable the proactive adaptation to a greater extent. This chapter discusses the design strategies that can be implemented to adapt to future climate conditions such as warmer temperatures and water shortages. The effect of extreme climate events (i.e. heat waves, flooding and storms) on buildings and strategies of resilient design for such events will also be discussed.

8.2 Sustainability

In the climate change domain, sustainable development can be considered as a solution on how current development may meet current demands without compromising the global climate in the future, which may eventually have a significant impact on, for instance, the welfare of future generations on the aspect of the way that human interacts with the biosphere to maintain its life support function on the aspect of biological diversity, ecosystem conservation and regional interconnectedness.

Climate change policies could be linked to the aspects of sustainable development (Agenda 21). Energy efficiency is an effective means to reduce greenhouse gas emission, while it is also in alignment with sustainable development. However, measures to meet mitigation and adaptation goals may not necessarily satisfy needs in all the economic, social and environmental dimensions of sustainable development. Measures that do not take all into account might not be sustainable in the long run.

© Springer Nature Switzerland AG 2021
X. Wang, S. Ramakrishnan, *Environmental Sustainability in Building Design and Construction*, https://doi.org/10.1007/978-3-030-76231-5_8

In general, while sustainable development may help to shape up climate change mitigation and adaptation policies, the integration of climate change mitigation and adaptation perspectives into built environment will make development more sustainable.

8.3 Built Environment

Built environment refers to the man-made surroundings that provide the setting for human activity, ranging from the building surroundings to the personal places. It addresses the relationship of human activities with building design, management and use of man-made surroundings.

Built environment, while it mostly considers issues related to building forms, is also a part of urban environment in social, economic and environmental domains. Building design and planning require the inclusion of urban environment context. Although infrastructure, public service and amenities may not necessarily fall in the immediate environment of building development, they are the critical elements for sustainable building development. Several measures, in terms of geospatial statistics, that may be considered are to quantify the building development:

Demography

- Population and population density
- Employment and employment density
- Family income and income per capita
- Education

Land use

- *balance*: Proportional balance of developed land uses, by (net or gross) land area, expressed on a scale of 0 (low) to 1 (high): 0.7–0.9 for well-balanced areas and 0.3–0.5 for imbalanced areas. It is described by $\sum P_i \ln P_i / \ln (N)$, where Pi is the proportion and N is the land uses. If all Pi are the same, the index is 1.
- *Space connectivity*: Proportion of space among a grid of cells of user-defined open space
- *mix*: Proportion of mixed or dissimilar developed land uses among a grid of cells of user-defined size, expressed on a scale of 0–1
- Open space and open space per capita

Housing

- Dwelling and dwelling density
- Amenity proximity: Average travel distance from all residents to closest designated amenity
- Transit proximity: Average travel distance from all residents to transit

- Amenity adjacency: Percentage of residents within a user-defined linear distance of user-designated amenities (e.g. school, community centre, shopping, etc.)
- Green space per dwellings

Public service

- Park space, hospitals and schools per dwellings
- Park, hospital and school proximity

Energy, water and waste

- Energy and water consumption
- Waste generation
- Energy and water consumption (or waste generation) per year per capita
- Energy and water consumption (or waste generation) per year per square meter
- Building forms are interwoven with environment or an element in the macro-urban context. While they are affected by environment, they also wield great impacts on the environment.

Impact of Environment on Buildings:

Urban salinity: Salt can affect buildings by varieties of ways: through the air, via the rain and through the ground. Salt attack through air and rain is generally classified as atmospheric corrosion. Salt attack through the soil in urban area is generally known as urban salinity.

Urban salinity is a major problem affecting the buildings. For instance, 68 towns in Australia have been identified as being affected by urban salinity. The salt and water transfer through the pores of building materials such as concrete, brick and stone causes the major damages to buildings, including efflorescence, deterioration of foundation and concrete slabs on ground and corrosion of underground services (Ashe et al., 2003).

The sulphate attack on the steel and concrete structures caused by acid sulphate soil is another problem affecting coastal areas from southern New South Wales (NSW) to northern Queensland. The presence of salt and acidity in acid sulphate soils lead to the risk of accelerated decay in Steel and concrete construction. There have been a number of developments at local and state levels in NSW, Victoria and WA to deal with the problem.

Pollution: Air Pollution: Exhaust gases from equipment are the main air pollutants from building. These include NO_x, SO_x and particulates (dust). Measures include the use of low-pollutant equipment, removing pollutants before discharge, positioning of the exhaust and the use of plants and other means for absorbing the pollutants.

Noise pollution: Noise pollution could come from a number of sources including the running of mechanical equipment, car movements, human activities and wind-induced noise.

Light pollution: Light pollution includes exterior lighting, light spill from interior, advertising display lighting and glare reflected from buildings.

Heat pollution: Wasted heat from equipment and heat radiated from structures and pavements could be considered as pollution to be eliminated if possible.

Impact of Buildings on Environment:

Apart from the impact of environment on a building during its lifecycle, the buildings are also impacting the environment in the following aspects (Fay, Vale, & Bannister, 2004).

Biodiversity: Issues include whether a survey has been made of the surrounding habitat before building, whether effort has been made to preserve existing ecology, the planting of appropriate vegetation, etc.

Greenhouse gas emissions: The greenhouse gas emissions caused during the construction, operation and demolition of buildings are a key factor in increasing the levels of carbon dioxide in the atmosphere, leading to anthropogenic climate change. The operation of buildings and occupant behaviour can affect their energy demand significantly.

Landscape: Considerations include whether the design of the building is responsive to the surrounding environment.

Water use: The building operation can be a major sector for water consumption. However, the building users can follow sustainable practices in harvesting water and reduce the water demands on existing constrained supplies.

Waste: The Waste is generated in all stages of buildings' lifecycle including construction, operation and demolition. The landfill wastes contributes to resource depletion and a range of pollutants and emissions. The effective reduction and recycle of waste minimises land contamination, and reduces the environmental impacts caused by waste materials.

Transport: Transport is a major source of Australia's greenhouse gas emissions. The transport sector related emissions are mainly caused by the location of buildings, where accessibility limits the possible transportation modes and encourages private transportation options, leading to high emissions.

Refrigerant use: Refrigerant use (e.g HFC and HCFC) in buildings is is a contributor to greenhouse emissions and ozone depletion. The environment safe refrigerant choices are critical to minimise the negative impact onto environment.

Storm water run-off: The built environment has altered the natural storm water and infiltration due to the increased impervious surfaces, leading to adverse impacts on marine life, and on freshwater environments. The sustainable storm water design by encouraging the pervious surfaces surrounding the buildings will minimise the disruption onto natural stormwater flows.

Sewage outfall volume: The volume of sewage sent out from buildings may exceed the capacity of water treatment facilities and increases the load on the existing sewage infrastructure, leading to the greater likelihood of overflows into the environment.

Sunlight obstruction: Impacts of shadows cast by a building on adjacent sites should be considered.

Airborne debris: Two types of airborne debris need to be considered: wind-driven airborne debris and human-induced airborne debris. Damage caused by the debris should be limited, particularly those that have the potential to cause injuries or threat to life.

8.4   Climate Change Implications to Buildings

Built environment is likely to endure greater exposure to climate changes, including the increase of temperature, precipitation and sea level as well as the variation in wind pattern, especially the change of intensity and frequency of extreme events related to heat wave, storm, tidal surge and cyclone along with their secondary effects in terms of drought, flood, flushing, subsidence, landslide and bush fire. The effects may be greatly exacerbated by combined climate variability and extreme events. The complex relationship between climate change and buildings is illustrated in Fig. 8.1. The consequence of anthropogenic climate drivers can be

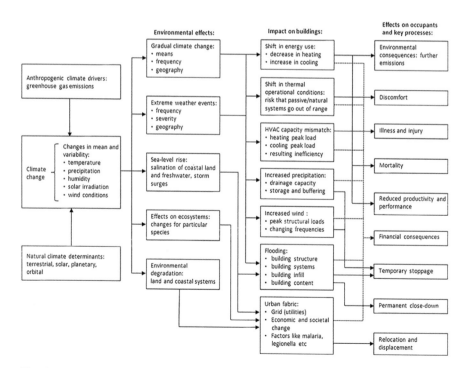

Fig. 8.1 Impact of climate change on environment, buildings and occupants. (Wilde and Coley 2012)

described as the environmental effects followed by the impacts on buildings and then the occupants.

Considering the buildings, the main risk from climate change to building structures is an increase in damage from specific events such as severe storms or cyclones, heat wave, drought, bush fire, etc. The fragility of building structures in resisting these events is generally non-linear (i.e. relatively little damage if the event is below the design benchmark but will be massive if above). For example, a 25% increase in wind speed above the design wind speed could lead to a fivefold increase in damage for buildings.

About 50% of buildings in Australia are within 7 km of the shoreline. These buildings, since located at or near the coast, they are potentially exposed to climate-related natural hazards such as floods, intense rainfalls, acid sulphate soils and tropical cyclones or extreme wind events. Other climate-related hazards such as bush fires, extreme temperatures and salinity are mainly affect the in-land buildings, however, also impact on coastal communities. Existing housing stock in Australia accounts for about 98% with 2% new stock to be added each year and a demolition rate of less than 0.5%. Thus, at least half of the dwellings in Australia by 2060 are yet to be constructed. The implications of the above are:

- Any action considered necessary to improve the resilience of the building stock needs to be taken as soon as possible for maximum effects.
- Measures are needed for both current existing stock and new stock.

A study conducted by BRANZ draws three broad conclusions about the resilience of Australia's new building stock to the likely impacts of climate change (Board, 2010):

- "New buildings are reasonably resilient to expected changes in average climate conditions but may not be as resilient to changes in extreme events such as storms and flooding.
- Some recent changes to building codes and practices, while not designed to address the impacts of climate change, have increased the resilience of new buildings. For example, higher energy efficiency standards mean that buildings are better able to cope with more frequent hot climate conditions.
- There is a considerable scope to improve the resilience of new buildings although further research may be required before specific measures can be formulated."

The resilience of Australia's older building stock to the likely impacts of climate change is difficult to assess, due to varying levels of degradation (corrosion, rot and insect attack) and a lack of information about the stock of buildings. However, a fundamental understanding of the changes in building's characteristics is necessary to understand for building refurbishment and new buildings for adapting to the changing climate. In this regard, this chapter will focus on three major impacts of climate change in buildings: thermal comfort, heat waves and durability of buildings.

8.5   Resilience for Thermal Comfort

8.5.1   *Thermal Comfort*

Thermal comfort refers to the person's state of mind that reflects satisfaction within the surrounding environment (ASHRAE, 2013). It is essential for occupants to work and perform duties within a comfortable thermal environment, although the satisfaction of comfort is different for each individual due to physiological and psychological variations. Occupants working in an environment with poor thermal comfort will feel discomfort, provide poor productivity and even suffer illness and fatigue.

Thermal comfort is influenced by heat conduction, convection, radiation and evaporative heat loss. It is maintained when heat, generated through human metabolism, dissipates from the body and maintains the thermal equilibrium within the surrounding environment. Any heat gain or loss beyond this will cause a sensation of discomfort. Based on these concepts, two models for the assessment of thermal comfort are generally adopted: Predictive mean vote (PMV) model and Adaptive model. ASHRAE (2004) also gave guideline to quantify indoor thermal comfort. The indicator of thermal comfort depends on six primary parameters that govern the conditions of thermal comfort. These parameters are:

- Rate of metabolism
- Insulation of clothing
- Air temperature
- Radiant temperature
- Air movement/velocity
- Relative humidity (RH)

All six parameters have a variation with time but are restricted to measuring thermal comfort in a steady-state environment only, where conditions are maintained with little variation. Because of this, occupants entering a particular environment may not feel initial comfort unless they have come from an environment with thermal comfort conditions similar to those in the new environment.

Predicted mean vote (PMV) is applied to measure thermal comfort applicable to an enclosure in which occupants perform normal routine office activities and the air speed is less than 0.2 m/s. It is not applicable in a physical outdoor environment involving occupants participating in sport. To define occupants' sensations of thermal comfort in this model, the indoor environment is rated as follows:

- +3 hot
- +2 warm
- +1 slightly warm
- 0 neutral
- −1 slightly cool
- −2 cool
- −3 cold

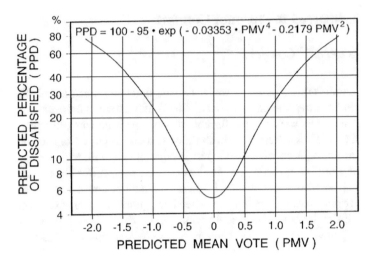

Fig. 8.2 Relation of PMV and PPD. (ASHRAE, 2013)

The above scales are obtained from calculation with the six parameters of thermal comfort based on heat balance principles. *Predicted percentage of dissatisfied* (PPD) is another term defined in this model and is used to describe the percentage of occupants that may feel uncomfortable in a given indoor environment. The relationship between PMV and PPD is shown in Fig. 8.2.

It is suggested that an acceptable indoor environment should be capable of maintaining both PMV and PPD in the range of $-0.5 < PMV < 0.5$; PPD $< 10\%$, which is shown in the shadowed zones in Fig. 8.3. It specified the comfort zone for environments that meet the above criteria and where the air speeds are not greater than 0.20 m/s . Two zones are shown, one for 0.5 clo of clothing insulation and one for 1.0 clo of insulation. These insulation levels are typical of clothing worn when the outdoor environment is warm and cool, respectively; *clo* is a unit used to express the thermal insulation provided by garments and clothing ensembles, where

$$clo = 0.155 \text{ m}^2 \text{ C} / \text{W}.$$

The PMV model can deal with variations in behaviour such as the changing of clothing and adjustment of air velocity. However, it does not handle aspects related to occupants, such as the changes and adjustments in occupants' personal thermal control in association with metabolism or clothing insulation, which is assumed to be constant for the PMV. Another thermal comfort measure, which entails the assumption that occupants can make adjustments themselves or to the thermal environment, was subsequently developed, known as the adaptive model.

The adaptive model was developed for an environment that is naturally ventilated with windows that are openable by the occupants. It combines the concept that the weather and changing seasons significantly influence people's behavioural adaptations and thermal sensations of the indoor environment. It also assumes that

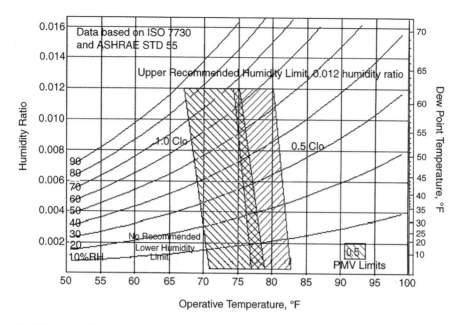

Fig. 8.3 Acceptable ranges of temperature and humidity in terms of PMV and PPD. (ASHRAE, 2004)

occupants are able to adjust their clothing or activities to adapt themselves to certain environments. Whereas the PMV model is ideal for mechanically ventilated enclosures, the adaptive model is an alternative for naturally ventilated spaces. Since this model was achieved from thousands of experiments and wide-range surveys around the world with different people, it is much simpler than the PMV model. Only two parameters are considered:

- Indoor temperature
- Mean monthly outdoor temperature

Figure 8.4 shows the rated comfort zones for the adaptive model.

8.5.2 Climate Change and Thermal Comfort

There is an increasing debate about the impact of changing climate on the indoor thermal comfort, particularly during summer times because future summers are likely to be both warmer and drier along with the increase in the occurrence of extreme temperatures. This will have a negative impact on the mechanically cooled buildings and free-running buildings, and the impacts are read as increasing energy consumption and prolonged thermal discomfort, respectively. Therefore, the owners and operators are keen to understand the resilience of existing building stock, and the clients of new buildings want to know how the proposed design will operate in

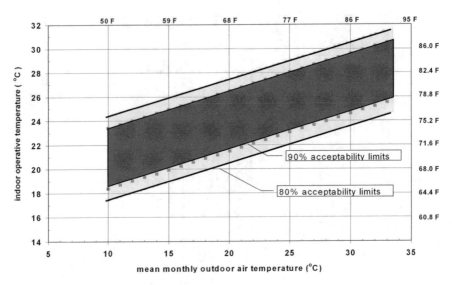

Fig. 8.4 Ranges of acceptable operative temperatures in terms of the adaptive model. (ASHRAE, 2004)

the future climate. What is required is a systematic and rational approach to identifying the buildings that are thermally susceptible to the changing climate. The thermal susceptibility of a building depends on its geographical location, the building type and function of the building and the vulnerability of the occupants to elevated temperatures. With the identification of these factors, the need for adaptation measures to the changing climate should be reliably predicted to undertake the necessary refurbishment measures and a sequence of appropriate interventions.

Location of the Building The geographical location of the building has an impact on the degree of adaptability to changing climate. For example, Australia has eight different climate zones, and the impact of climate change significantly varies on these climate zones (Fig. 8.5). A study conducted by Wang et al. (2010a) on the impact of climate change on the heating/cooling energy demands of Australian residential buildings revealed that the hot or warm climate zone (Darwin and Alice Springs) and warm temperate climate zone (Sydney) has high cooling energy demands than cool temperate climate (Hobart) and mild temperate climate (i.e. Melbourne). This is because the cool and mild temperate climate zones have a considerable reduction in the winter heating energy due to warming climate, and the total energy increase would not be significant.

Building Type and Function of the Building The building type and its function have an impact on the adaptation and resilience to the changing climate. For instance, the residential buildings and office buildings can adopt to a large range of climate change impacts compared with the hospitals, age-care centres, health care

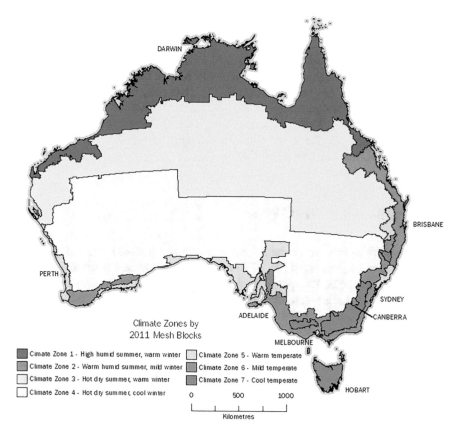

Fig. 8.5 Climate zones in Australia. (Peter Lyons et al., 2013)

organisations, etc. because the latter building types are expected to provide a safe haven for those at large who are suffering during the thermally uncomfortable periods such as heat waves. As these places containing a large group of sick and vulnerable people and therefore the health concerns of these people should be a high priority. A study conducted on the free-running hospital buildings revealed that the hospital wards deemed uncomfortable to the predicted climate in 2080 for the adaptive thermal comfort models. However, these buildings can improve the thermal comfort if some passive cooling techniques, such as using fans, in ward buildings are used.

Another example is the gym and indoor sport venues. These buildings generally have regulations of nonoperation when the outdoor temperature exceeds 35 °C. As the warming climate evidences increasing number of days experiencing this upper bound, the functions of these buildings are significantly affected. In addition, these buildings do not have mechanical cooling systems and generally rely on the passive cooling methods. Therefore, the building type and its function should be given a priority when assessing its resilience to changing climate.

Vulnerable Population in the Building The vulnerable population includes older adults above 65 and infants to the very young kids up to the age of 6 years who are more susceptible to the changing climate. In particular, the extreme heat wave events have caused a significant risk of mortality and morbidity in this age range (Santamouris, 2016). While the heat waves have a serious effect on non-air-conditioned buildings, the air-conditioned buildings are also affected by power blackouts and brownouts and put at risk.

8.6   Resilience to Heat Waves

The heat wave events are characterised by prolonged high-temperature exposure and have caused an increased risk of mortality and morbidity. For instance, during July and August 2003, European countries experienced its worst ever heat wave, where the temperatures recorded above 40 °C. In 16 European countries, more than 70,000 excess deaths occurred during the summer of 2003 (Cadot et al., 2007). France was particularly affected, where 15,000 excess deaths were observed. The greatest increase in mortality was due to heat-related issues such as dehydration, hyperthermia and heat stroke (Fouillet et al., 2006). The US states have also experienced exceptional heat waves in 1995 and 2006 causing 692 and 505 excess deaths, respectively (Kaiser et al., 2007; Ostro et al., 2009). In 2011 summer, Houston and other areas in Texas experienced an exceptional heat wave producing the hottest August on record. However, no significant excess deaths were reported during this period. The possible reasons were reported as high prevalence of air-conditioning and acclimation to heat among residents. Heat-related deaths have been shown to decrease with increased air-conditioning prevalence (O'Neill et al., 2005).

The impact of heat waves on buildings and occupants has also been reported in Australia. In February 2004, almost two-thirds of continental Australia recorded the temperatures above 39 °C. Brisbane in particular recorded 41 °C and 42 °C for two consecutive days of 21 and 22 February 2004. This heat wave event caused 116 additional deaths in Brisbane, Australia (Ren et al., 2014; Tong et al., 2010). During January 2009, south-eastern Australia (i.e. South Australia and Victoria) experienced an extreme heat wave event. During the period of 27–31 January, the majority of the Victorian weather stations recorded maximum temperatures, more than 12–15 °C above normal. On 30 January, the maximum temperature in Melbourne was recorded as 45.1 °C, which is the second highest on record. This heat wave event caused 374 additional deaths in Victoria (Nguyen et al., 2010; Ren et al., 2014).

As summarised above, heat wave events can have life-threatening consequences with increased temperatures and reduced diurnal temperatures within the built environment and needing a substantial response from emergency services (Vandentorren et al., 2004). It was stated that, in the absence of any modification to current climate change trend, the summer observed in Paris during 2003 is predicted to be typical and average in 2040. This will clearly have a significant impact on the energy consumption of air-conditioned buildings and the thermal comfort of naturally controlled buildings. Coley and Kershaw (2010) evaluated the relationship of future climate conditions with the indoor thermal conditions and found that the relationship is linear among all studied building types. They introduced a term to understand the proportional change in indoor temperature condition to outdoor future climate as *climate change amplification coefficients*. The main outcomes of this study can be summarised as follows:

- The response of indoor thermal conditions for the changing climate is linear, regardless of the architecture, construction, ventilation type or use of the building.
- The gradient of the change is different for different building types.
- The gradient of the linear relationship is greater than unity for some buildings, while others show less than unity. This means the former building type amplifies the effects of climate change, while the latter suppresses the effect of climate change.
- There is no need to do multiple simulations for different climate scenarios, two simulations are enough to identify the gradient, and response of the building for other scenarios can simply be calculated from the gradient.

Figure 8.6 shows the gradient or trend of the response of indoor environment to changing climate. The estimation and use of these coefficients for new or existing

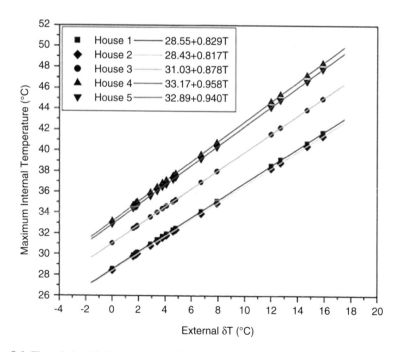

Fig. 8.6 The relationship between external temperature and internal temperature in passively cooled buildings. (Coley & Kershaw, 2010)

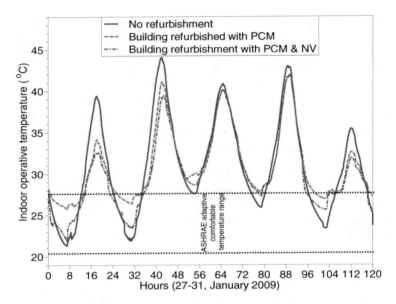

Fig. 8.7 Indoor temperature variation for various refurbishment methods during the heat waves in Melbourne, Australia. (Ramakrishnan et al., 2017)

buildings will enable multiple benefits including the design of more resilient buildings adapted to a changing climate, cost–benefit analysis of refurbishment options and the rational assembly of at-risk registers of vulnerable building occupants.

Studies were also attempted to understand the influence of efficient building design and refurbishment measures that can reduce the effect of heat waves by sheltering occupants and reducing the heat stress risks indoors. A study by Ramakrishnan et al. (2017) reported that the thermal energy storage enhancement of buildings using advanced thermal energy storage systems such as phase change materials will reduce the heat stress during extreme heat waves. Authors conducted a simulation study using the 2009 Melbourne heat wave event and reported the indoor operative temperature of a standard single-storey building for different cases of no refurbishment, building refurbished with phase change materials and building refurbishment with phase change materials and night ventilation. The effect of phase change material and night ventilation is clearly revealed as given in Fig. 8.7 below.

8.7 Resilience for Durability

8.7.1 Building Durability

Durability is normally referred as the ability of materials and structural components as well as buildings to maintain their integrity over time. For buildings as a whole, it is considered as the ability to provide occupants safety and serviceability over

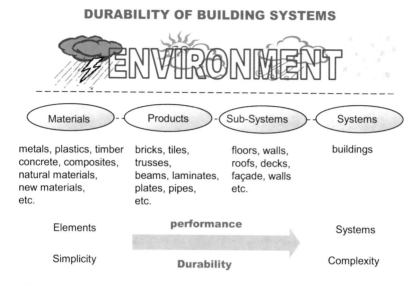

Fig. 8.8 Durability of building systems at multiscale

time. The assessment of durability of a whole building can be challenging, and quite often, the durability is assessed based on the scale at which failure to meet requirements occurs. For instance, some durability failure occurs at a material level of the building components, whereas other failures could be a component/product or even a subsystem failure. The below definitions identify the different durability-based failure modes that can be experienced in buildings. Figure 8.8 also shows how the different-scale assessment can be related to the other scales.

Durability at Different Scales

- Material scale: Physical properties such as mechanical strength and resistance to degradation that caused corrosion for steel, decay for timber, carbonation, acid and sulphate attack and alkali–silica reaction for concrete.
- Product scale: Geometrical characters which directly affect mechanical and service performance, such as tensile, compressive and bending force and buckling force threshold and resistance to water or moisture penetration
- Subsystem and system scale: Interconnection among products or components, which may affect the integrity and serviceability of building systems

The impact of climate change can also be seen on the long-term performance of building materials. In particular, the commonly used construction materials of concrete and timber are highly sensitive to the environmental conditions, and the changes in these conditions can impact the performance of these materials. One of the key performance criteria of building and infrastructures is to provide adequate safety, serviceability and durability. The decision-making related to the development of buildings and infrastructure should take into account the variation in the above requirements for the future climate conditions. The study of the implications of climate change on the protection of building and infrastructure is crucial for effective decision-making.

8.7.2 Deterioration of Concrete Structures Under Changing Climate

The deterioration of concrete or reinforced concrete can be caused by physical factors – freeze–thaw cycles, thermal mismatch between cements and aggregates; mechanical factors – erosion, scaling and abrasion; and chemical factors - penetration of chemicals from the environment, such as atmospheric CO_2 and reactive ions. As the atmospheric CO_2 penetrates into concrete, carbonation occurs, leading to the reduction of the pH and increasing the vulnerability of steel reinforcement to depassivation and the loss of corrosion protection. Corrosion products cause considerable expansion, generating internal stress and causing cracking, spalling or delamination. Chloride-induced corrosion is the major threat to marine and coastal infrastructures.

Concrete structures show a responsive effect to the variation in atmospheric carbon dioxide concentration, temperature and relative humidity. The dramatic increase in the atmospheric CO_2 concentration from 280 ppm in pre-industrialisation age to the 400 ppm in 2020 can have a significant negative implication to the reinforced concrete structures. On the other hand, the increase in global temperature and change in humidity can also affect the concrete structures by accelerating the concrete deterioration processes. The consequences due to the variation in the above-mentioned factors are studied by Wang et al. (2010b) and listed in Table 8.1 below.

8.7.3 Prevention of Concrete Deterioration

The deterioration rate of concrete structures heavily depends on the environmental exposure conditions. For instance, in Australia, the environmental exposure is classified on the basis of macroclimate as arid, temperate and tropical zones as depicted in Fig. 8.9 below.

This has resulted in the new classification of exposure conditions as A1, A2, B1, B2, C1 and C2 according to the severity of deterioration. Therefore, design, operation and maintenance of concrete structures should be aligned with their exposure conditions that characterise the minimum strength requirements, cover to reinforcement, water/cement ratio and other related properties to maintain appropriate durability. For new constructions, the increase in concrete cover, strength increase and using stainless steel reinforcement are most widely used solutions to overcome the deterioration. Meanwhile, existing structures should also be maintained to minimise

Table 8.1 Factors and implications of concrete deterioration (Wang et al., 2010b)

Factor	Implications
Increase of CO_2 concentration	Elevated carbon concentration accelerates carbonation and increases carbonation depth in concrete: This increases the likelihood of concrete structures exposed to carbonation-induced reinforcement corrosion initiation and structural damage
Change of temperature	Elevated temperature accelerates carbonation, chloride penetration and corrosion rate of reinforcement that exacerbates the corrosion damage
Change of humidity	Lowered relative humidity may reduce or even stop carbonation and chloride penetration in the area with yearly average RH currently just above 40–50%, while increased humidity may result in them occurring in the regions where they are now negligible.

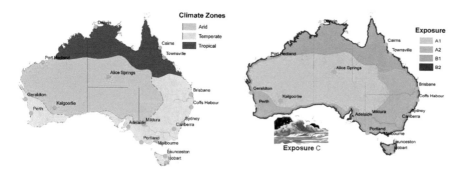

Fig. 8.9 Classification of exposure conditions in Australia. (Wang et al., 2010b)

the deterioration, and the possible solutions are the creation of exposure barriers, preventing the penetration of aggressive substances and removal and replacement of deteriorated parts to prevent further damages. Some of the currently used mitigation methods of concrete deterioration are presented in Fig. 8.10 below.

8.7.4   Durability of Timber

Timber is widely used as structural and nonstructural building materials in residential and low-rise commercial building applications. Furthermore, based on the contact conditions, timber can be classified as in-ground and above-ground application methods. Being as bio-based material, the durability of timber is affected by microbial and weathering that are major concerns. The different hazards on the timber can be listed as below:

- Fungal attack
  - In-ground
  - Above-ground

Option 1: Creation of exposure barriers
  o Moisture barriers, such as waterproofing
  o Protective coatings for additional protection: e.g. epoxy (non-breathable moisture barrier), polyesters, acrylics (which allow water diffusion), polyurethane, bitumen, copolymer, and anti-carbonation coating (acrylic materials)
  o Coatings on steel reinforcement
  o Surface preparation of concrete and reinforcement

Option 2: Preventing the penetration of deleterious substances
  o Polymer impregnation, such as percolating into concrete substrate

Option 3: Extraction of deleterious substances
  o Cathodic protection that migrates chloride ions from the steel surface towards an anode
  o Chloride extraction, by which chlorides are transported out of the concrete to an anode surface
  o Re-alkalization by applying an external anode to the concrete surface (with the steel reinforcement inside the concrete acting as the cathode) and an electrolytic paste (comprising sprayed cellulosic fibre in a solution of potassium and sodium carbonate); the electrolyte moves into the concrete, increasing alkalinity

Option 4: Removal and replacement of deteriorated parts of structures
  o Patch repair systems, such as the renewal and/or preservation of the passivity of steel reinforcement and the restoration of structural integrity by applying mortar or concrete to areas where deterioration occurs; materials for patch repair can be cementitious or epoxy-based, or comprise similar resinous materials (e.g. polymer concrete and polymer-modified concrete)
  o Concrete removal: e.g. the removal of damaged or deteriorated areas

Fig. 8.10 Options to mitigate concrete deterioration. (Wang et al., 2010b)

- Insect attack

  – Termites and borers
- Corrosion of fasteners
- Weathering
- Marine borers
- Chemical degradation
- Fire

AS1064 specifies the natural durability, preservative treatments and hazard levels of timber for various applications. The decay of timber can be favourable in the following conditions:

- The appropriate moisture must be present: Moisture content of 0–20% is too dry – attack will not occur, 20–60% – sufficient moisture for attack to occur, >60% is too wet with insufficient oxygen for attack to occur.
- Oxygen must be present. Timber completely submerged or saturated is rarely attacked, and timber below ground for 600 mm or more is also rarely attacked due to the lack of available oxygen.

- Temperature must be in the favourable range of 5–40 °C for timber to decay. 25–40 °C is ideal. At lower temperatures, fungal attack is retarded. At higher temperatures, the fungus will not survive.
- Food in the form of nutrients (carbohydrates, nitrogen, minerals, etc.) must be present. These are usually provided by the timber itself, particularly sapwood that is normally high in sugars and carbohydrates.

The protection measures for timber from fungal attack can be achieved:

- By eliminating contact with moisture
- By using species with a durability appropriate to the application or by using species that have been preservative treated to a level appropriate to the hazard

The AS5604 specifies the probable lifespan of untreated timber for various exposure conditions. However, these typical values may be varied due to natural variability of durability or the variation in hazard levels (Table 8.2).

8.7.5 Deterioration of Timber Under Changing Climate

The future climate projections in the UK suggest that the winter will experience warmer and wetter conditions and the summer will have warmer and dryer days. This change in climate condition can be favourable to the microbiological organisms including fungi, bacteria and insects and accelerates the attack on bio-based materials such as timber. The decay in wood, particularly the exterior timber that is in-ground, can cause a significant economic loss along with high maintenance costs and reduced lifetime (Curling & Ormondroyd, 2020).

A study on the changing climate condition in Australia on the timber decay has suggested that, under different emission scenarios, the natural durability of wood species generally has a negligible effect on the changes of decay rate. Under the three emission scenarios, the median decay rate of wood by 2080, relative to that in 2010, could increase up to 10% in Brisbane and Sydney, but could decrease by 12%

Table 8.2 Timber classes and life expectancy in Australia (Australia, 2005)

Natural durability class	Probable heartwood life expectancy (years)		
	Fully protected from the weather and termites	Above ground exposed to the weather but protected from termites	In-ground contact and exposed to termites
Class 1 highly durable	50+	40+	25+
Class 2 durable	50+	15–40	15–25
Class 3 moderately durable	50+	7–15	5–15
Class 4 nondurable	50+	0–7	0–5

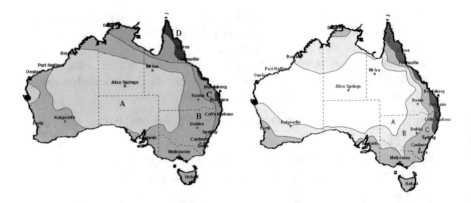

Fig. 8.11 Hazard maps – timber decay (A: lowest decay hazard zone; D: highest decay hazard zone). (Wang et al., 2007)

in Melbourne. However, the residual strength of timber has been significantly dropped. For less durable wood materials, the residual strength after 50 years of installation has been dropped by 25% due to climate change, compared with without climate change (Wang & Wang, 2012). The contradicting behaviour of decay in the Australian cities can be correlated to the temperature and rainfall data. For Brisbane and Sydney, the upward trends of temperature dominate the decay-slowing effect compared with the decay-accelerating effect of the slightly downward trends of rainfall. For Melbourne, the decreasing rainfall trend has a favourable effect more than compensating for the reduction in decay due to increasing temperature (Fig. 8.11).

8.8   Summary

This chapter explained how climate change affects the operating performance of buildings in the future. The major threats of climate change to the buildings are increasing air temperature, reducing the relative humidity, increasing the CO_2 concentration in the air and the occurrence of extreme events. The resilience of the building is assessed under three subjects of resilience to thermal comfort, resilience to heat waves and resilience to durability of construction materials. There are other building resilience concerns with the climate change including the resilience to precipitation changes (flash flooding), resilience to sea level rise and disaster management to events like storms, drought, cyclones, etc. However, they are beyond the discussion of the current book. Nevertheless, this chapter provided an insight of how building and environment are interconnected and the changes in one will reflect on the other.

References

Ashe, B., Newton, P. W., Enker, R., Bell, J., Apelt, R., Hough, R., Davis, M. (2003). Sustainability and the building code of Australia.

ASHRAE, A. (2013). Standard 55-2013: "Thermal environmental conditions for human occupancy"; ASHRAE. In *Atlanta USA*.

Australia, S. (2005). Timber—Natural durability ratings. In *AS 5604*. Australia.

Board, A. B. C. (2010). An Investigation of Possible Building Code of Australia (BCA)

Cadot, E., Rodwin, V. G., & Spira, A. (2007). In the heat of the summer. *Journal of Urban Health, 84*(4), 466–468.

Coley, D., & Kershaw, T. (2010). Changes in internal temperatures within the built environment as a response to a changing climate. *Building and Environment, 45*(1), 89–93. https://doi.org/10.1016/j.buildenv.2009.05.009

Curling, S. F., & Ormondroyd, G. A. (2020). Observed and projected changes in the climate based decay hazard of timber in the United Kingdom. *Scientific Reports, 10*(1), 16287. https://doi.org/10.1038/s41598-020-73239-1

Fay, R., Vale, R., & Bannister, P. (2004). The National Australian Built Environment Rating System (NABERS). Environment Design Guide, 1–6.

Fouillet, A., Rey, G., Laurent, F., Pavillon, G., Bellec, S., Guihenneuc-Jouyaux, C., … Hémon, D. (2006). Excess mortality related to the August 2003 heat wave in France. *International Archives of Occupational and Environmental Health, 80*(1), 16–24.

Kaiser, R., Le Tertre, A., Schwartz, J., Gotway, C. A., Daley, W. R., & Rubin, C. H. (2007). The effect of the 1995 heat wave in Chicago on all-cause and cause-specific mortality. *American Journal of Public Health, 97*(Supplement_1), S158–S162.

Nguyen, M., Wang, X., Chen, D., & Flagship, C. A. (2010). *An investigation of extreme heatwave events and their effects on building & infrastructure*. National Research Flagships Climate Adaptation.

O'Neill, M. S., Zanobetti, A., & Schwartz, J. (2005). Disparities by race in heat-related mortality in four US cities: The role of air conditioning prevalence. *Journal of Urban Health, 82*(2), 191–197.

Ostro, B. D., Roth, L. A., Green, R. S., & Basu, R. (2009). Estimating the mortality effect of the July 2006 California heat wave. *Environmental Research, 109*(5), 614–619. https://doi.org/10.1016/j.envres.2009.03.010

Peter Lyons, B., Chris Reardon, C., & Tracey Gramlick, R. (2013). *Passive design, glazing—Your home: Australia's guide to environmentally sustainable homes*. Commonwealth of Australia (Department of Industry).

Pieter de Wilde, David Coley, (2012) The implications of a changing climate for buildings. Building and Environment 55:1–7

Ramakrishnan, S., Wang, X., Sanjayan, J., & Wilson, J. (2017). Thermal performance of buildings integrated with phase change materials to reduce heat stress risks during extreme heatwave events. *Applied Energy, 194*, 410–421. https://doi.org/10.1016/j.apenergy.2016.04.084

Ren, Z., Wang, X., & Chen, D. (2014). Heat stress within energy efficient dwellings in Australia. *Architectural Science Review* (ahead-of-print), 1–10.

Santamouris, M. (2016). Innovating to zero the building sector in Europe: Minimising the energy consumption, eradication of the energy poverty and mitigating the local climate change. *Solar Energy, 128*, 61–94. https://doi.org/10.1016/j.solener.2016.01.021

Tong, S., Ren, C., & Becker, N. (2010). Excess deaths during the 2004 heatwave in Brisbane, Australia. *International Journal of Biometeorology, 54*(4), 393–400.

Vandentorren, S., Suzan, F., Medina, S., Pascal, M., Maulpoix, A., Cohen, J.-C., & Ledrans, M. (2004). Mortality in 13 French cities during the August 2003 heat wave. *American Journal of Public Health, 94*(9), 1518–1520.

Wang, C.-h., & Wang, X. (2012). Vulnerability of timber in ground contact to fungal decay under climate change. *Climatic Change, 115*(3–4), 777–794.

Wang, C., Leicester, R., Foliente, G., & Nguyen, M. (2007). *Timber service life design guide.* Forest and Wood Products Australia Limited. www.fwpa.com.au

Wang, X., Chen, D., & Ren, Z. (2010a). Assessment of climate change impact on residential building heating and cooling energy requirement in Australia. *Building and Environment, 45*(7), 1663–1682. https://doi.org/10.1016/j.buildenv.2010.01.022

Wang, X., Nguyen, M., Stewart, M., Syme, M., & Leitch, A. (2010b). *Analysis of climate change impacts on the deterioration of concrete infrastructure–synthesis report.* Published by CSIRO, Canberra. ISBN978 0, 643(10364), 1.

Chapter 9
Summary and Conclusions

This book provides an in-depth understanding of sustainable development in the construction sector and provided information on concepts, issues and implementation methods to achieve sustainable construction. The book was written to help graduate students, teachers and engineers to apply the knowledge in building and construction practices and making more informed decisions for the application of specific strategies suitable for the requirements. The sustainable solutions for building construction can vary significantly as some solution may assist during the construction stage, however, detrimental to the operation stage and vice versa. For example, a specific insulation material can perform well during the operation stage, but the embodied energy of making the insulation can be very high. In these situations, the engineers should be able to justify their decision-making considering the life cycle assessment of the building. The primary topics discussed in this book can be outlined as follows:

1. Definitions of sustainability, sustainable development, sustainable construction and related topics and challenges in implementing sustainability practices
2. The international and national policy developments to define a targeted sustainable development with the key indicators of those policies
3. The definitions of climate change terms, global climate observations and climate projection methods using various emission scenarios. The coping methods to climate change, such as mitigation and adaptation, are also discussed
4. Energy and carbon accounting in buildings throughout the life cycle
5. The resource efficiency in buildings with how materials and water can be efficiently used for building construction and in operation stages
6. The sustainable solid waste management in buildings with proper waste treatment methods and waste to resource management techniques
7. Sustainable building design opportunities and strategies for achieving green building design, low-energy design and zero-energy design
8. Resilience of building to changing climate with the resilient measures of thermal comfort, heat waves and durability of building materials

© Springer Nature Switzerland AG 2021
X. Wang, S. Ramakrishnan, *Environmental Sustainability in Building Design
and Construction*, https://doi.org/10.1007/978-3-030-76231-5_9

These topics were covered with examples and suggestions for better understanding of different strategies that enable the readers to think out-of-the-box for developing new technologies. The assessment methods and critical evaluation techniques can be very useful to explore materials and processes in construction projects that are more sustainable and/or require less embodied energy with the reduced level of greenhouse gas emissions.

The awareness on climate change in terms of how construction practices contribute to the climate change as well as the buildings' response to climate change is discussed. While the former issue explains how construction practices can be performed in a sustainable way to reduce the climate change contribution, the latter issue addresses how buildings should be designed and constructed to respond in a resilient manner for future climate conditions. The understanding on the global climate change observations and future climate projection methods can be very useful to find the interaction between buildings and climate.

Sustainable resource management is very critical in the construction sector as they use a vast number of resources for construction. There are six primary themes of land, materials, water, waste, energy and indoor environment were discussed. The efficient resource management of materials, water and waste management were discussed in detail.

The life cycle approach can be applied to many parameters including energy, emission, cost, etc., and the discussion on these topics is addressed at many places as this method is effective in quantifying the direct and indirect impacts of decision-making. The zero-energy/zero-emission design relies on life cycle approach and provides an insight of how each design/construction method can impact the life cycle energy/emission.

Finally, the buildings and infrastructures are expected to be operated for a long lifespan of 50–100 years, and the resilience of buildings to changing climate should not be ignored. The major vulnerability on buildings' performance was reported as the effect on indoor thermal comfort, heat wave effects and durability of building materials. The adaptation of buildings to climate change should be considered in design and construction stage in addition to the mitigation technologies, which are addressed through sustainable development methods.

Index

© Springer Nature Switzerland AG 2021
X. Wang, S. Ramakrishnan, *Environmental Sustainability in Building Design
and Construction*, https://doi.org/10.1007/978-3-030-76231-5

169

Printed in the United States
by Baker & Taylor Publisher Services